Profitable Partnering
for
Lean Construction

Profitable Partnering for Lean Construction

Clive Thomas Cain CBE
Building Down Barriers
UK

Blackwell
Publishing

© 2004 by Clive Thomas Cain

Editorial offices:
Blackwell Publishing Ltd, 9600 Garsington Road, Oxford OX4 2DQ, UK
   Tel: +44 (0)1865 776868
Blackwell Publishing Inc., 350 Main Street, Malden, MA 02148-5020, USA
   Tel: +1 781 388 8250
Blackwell Publishing Asia Pty Ltd, 550 Swanston Street, Carlton, Victoria 3053, Australia
   Tel: +61 (0)3 8359 1011

The right of the Author to be identified as the Author of this Work has been asserted in accordance with the Copyright, Designs and Patents Act 1988.

First published 2004 by Blackwell Publishing Ltd

Library of Congress Cataloging-in-Publication Data
Cain, Clive Thomas.
      Profitable partnering for lean construction/Clive Thomas Cain. – 1st ed.
      p. cm.
   Includes bibliographical references and index.
   ISBN 1-4051-1086-4 (pbk. : alk. paper)
      1. Building–Cost control. 2. Partnership. 3. Building materials. 4. Delivery of goods. I. Title.

TH438.15.C35 2004
690′.068′1–dc22

                                                              2004001469

ISBN 1-4051-1086-4

A catalogue record for this title is available from the British Library

Set in 12 on 14 pt Souvenir
by Kolam Information Services Pvt. Ltd, Pondicherry, India
Printed and bound in India
by Replika Press Pvt. Ltd, Kundli

The publisher's policy is to use permanent paper from mills that operate a sustainable forestry policy, and which has been manufactured from pulp processed using acid-free and elementary chlorine-free practices. Furthermore, the publisher ensures that the text paper and cover board used have met acceptable environmental accreditation standards.

For further information visit our website:
www.thatconstructionsite.com

Contents

Introduction

To understand how partnering ought to be used to drive forward radically improved performance in the construction sector, we need to do as the two UK Egan reports, *Rethinking Construction* and *Accelerating Change*, advocate. We need to look to other business sectors and learn from their experiences over recent decades.

In the manufacturing sector, improved performance relates primarily to the elimination of unnecessary costs (in terms of the inefficient utilisation of labour and materials down through the entire supply chain). The manufacturing sector's long experience of lean thinking has taught that this can only be done with any real success if all the firms in the supply chain collaborate closely together within long-term, strategic supply-side partnerships. In the UK aerospace sector, these long-term, strategic partnerships created an entity that became known as a 'virtual company' and their purpose is to improve the competitiveness of all the firms in the virtual company by enabling them to convert unnecessary costs into higher and more assured profits and lower prices for the demand-side customer, whilst improving the technical quality of the goods they produce.

If the construction sector is to replicate the performance improvements of other business sectors, it needs to employ supply chain management and lean thinking techniques like partnering in the same way as those other sectors. Thus the construction sector needs to see partnering primarily as a supply-side tool that is a fundamental part of the lean thinking aspect of effective supply chain management and is the mechanism whereby unnecessary costs are converted into higher profits and lower prices.

This divergence of understanding between the construction sector and all other sectors about the true meaning of the term partnering was picked up in a recent book by Mike Murray and David Langford called *Construction Reports 1944–98*. They reminded us that in a leading article in *Building* magazine in March 1999 it had been argued strongly that:

'The most famous buzzword of all, partnering, has been subject to a lot of abuse. It has been highjacked by consultants and corrupted by contractors'.

If the construction sector is to learn from other sectors it needs to understand that supply-side partnering is the mechanism that enables lean thinking to flourish. Partnering should be about ending the selection of sub-contractors and suppliers by lowest price competition for each contract (which invariably causes the make up of the design and construction team to constantly change from project to project and prevents any real and continuous improvement of the processes or of the constructed product). Partnering should provide supply-side firms with a culture of security and trust that encourages everyone to objectively and accurately measure performance and share the result without fear of criticism or risk of rejection, and to then work together to improve each other's performance.

Partnering should be about creating a virtual company of supply-side design and construction firms that is able to compete more effectively in its chosen market because its constructed products are of far better quality and offer far better value for money than those of its competitors. At the same time, partnering should ensure far better and more assured profits for every member of the virtual company than can be achieved by those firms that have not embraced long-term, strategic supply-side partnering.

The book is written in straightforward language that ensures ease of use by those at the sharp end of each sector of the construction industry. It endeavours to end the confusion about the meaning of buzzwords, such as partnering, by giving explanations and definitions that are drawn directly from best practice publications that are in current use in other business sectors.

The book explains why long-term, strategic supply-side partnering will deliver radically better value for money and far better profits. It examines the wide variety and size of end-user customers and then explains why partnering is impractical for the small and occasional end-user customers that form the bulk of the construction industry's market. It also explains why supply chain management and lean thinking tools and techniques mean that partnering with the demand-side customer is of secondary importance to supply-side partnering.

It explains why end-user customers of constructed products have been demanding radical improvement from the construction industry for at least the last 70 years and it explains why long-term, strategic supply-side partnering is a prerequisite of radical improvement. It also compares the demand for construction industry reform in the UK with the demand for reform in other major countries, such as the USA, Canada and Australia.

The book explains the concept of a virtual company and describes the long-term, strategic, supply-side partnerships that are essential for its effective operation. It explains how strategic supply-side partnering can improve the profitability and the competitiveness of small and medium enterprises as well as that of large firms. It describes the seven principles of supply chain management that are the foundation of successful and profitable supply-side partnering in every sector.

The book also explains why demand-side clients who are intent on improving the value for money they get from the construction industry should deal with the construction industry in the same way that they deal with all other business sectors, and should insist on hard evidence of long-term, strategic supply-side partnering before awarding construction contracts (or consultancies). It then goes on to suggest a possible selection process for demand-side clients.

Acknowledgements

The author wishes to acknowledge the help and advice he received from Iain Beaton, who developed a very successful best value tendering system whilst he was the Deputy Chief Executive of St Helens Metropolitan Borough Council. Without the benefit of Iain's help, Chapter 7 on the client's selection process would have been less relevant to the needs of client organisations.

1   What is Partnering?

Whenever I read about partnering in construction industry publications, or whenever I come across a discussion on partnering within the construction industry, I am struck by the widely diverging views of what is meant by the term 'partnering'. I am also struck by the fact that none of these views match what partnering is understood to mean, or how partnering is practised in other business sectors.

Highly successful companies in other business sectors have a remarkably common and unified understanding of what is meant by partnering. They have a shared view of where within the supply chain partnering delivers the greatest benefit, how partnering ties in with effective supply chain management and lean techniques, and how partnering ties in with the elimination of unnecessary costs and the delivery of high quality.

I am not alone in my concerns about the construction industry's misuse of the term 'partnering'. In a leading article in *Building* magazine in March 1999, Andrew Sims said:

'*The most famous buzzword of all, partnering, has been subject to a lot of abuse. It has been hijacked by consultants and corrupted by contractors.*'

If the construction industry is to move forward with a programme of radical reform and improvement, it needs to first
agree what is meant by the many buzzwords that fly around
the industry. In my earlier books, *Building Down Barriers –
A guide to construction best practice* and *Performance
Measurement for Construction Profitability* (see Further
Reading), I endeavoured to explain and clarify precisely what
is meant by terms such as 'lean construction', 'supply chain
management', 'integration' and 'performance measurement'. Whilst the two books touched on the meaning of
the term 'partnering', this book provides a detailed guide
for those firms that want to walk the talk. In Chapter 9 I have
included explanations and definitions of the many buzzwords
in common use within the construction industry by cross-
reference to well established definitions from other business
sectors (thus avoiding any accusations of bias).

In all other business sectors, partnering is a concept that
has been practised very effectively for several decades and
has been the foundation of the improvements in quality,
price and profitability. The understanding of what is meant
by the term partnering is common to all non-construction
sectors, as are the practices and the processes of partnering.
It is well understood that partnering is the primary mechanism by which supply-side firms drive out unnecessary costs
and drive up quality. Whilst they recognise that partnering
can occasionally occur between demand-side customers and
supply-side companies, they understand that partnering is
 primarily an activity that is confined to the supply-side of the
industry and must be in position long before any partnering
relationships with demand-side customers are considered.
They see it as the key to open, constructive and productive
relationships between the firms that form the supply chain
for any given product or service.

In business sectors such as aerospace or electronics, end
suppliers develop strategic partnerships with a number of
firms in each supplier category. These long-term strategic
partnerships are exclusively supply-side and do not involve

the demand-side customers of the manufactured product. They are put in place to ensure that the entire manufacturing supply chain can compete more effectively by mobilising the key members of the supply chain to work closely together to offer lower prices, better quality (in terms of the demand-side customers' needs), and better, more predictable profit margins. Figure 1.1 illustrates these supply-side partnerships and shows the interface between the demand-side customers and the end supplier.

As Figure 1.1 shows, although the demand-side customers in other business sectors interface briefly with the end supplier to purchase the goods manufactured by the end

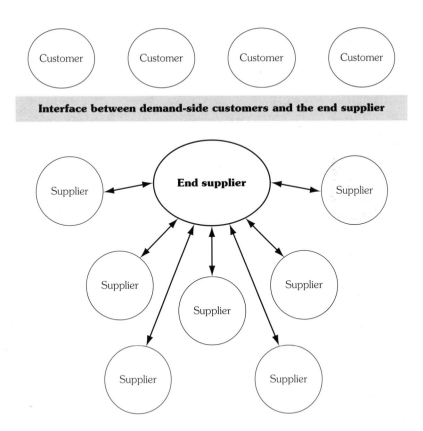

Figure 1.1   Partnering in non-construction business sectors

supplier, they very rarely form an intrinsic part of the long-term, strategic relationships that link together the firms in the end supplier's supply chain. In those other business sectors, partnering is a well developed and well proven long-term supply-side mechanism to facilitate continuous improvement on a year-by-year basis. The customer's role in such supply-side partnerships is to provide feedback, in the form of input into customer surveys, to the supply-side virtual company (the end supplier and the supply chain firms) on the quality and value for money of the goods purchased from the virtual company.

As was stated in *Building Down Barriers Handbook of Supply Chain Management (see Further Reading):*

'The commercial core of supply chain integration is setting up long-term relationships based on improving the value of what the supply chain delivers, improving quality and reducing underlying costs through taking out waste and inefficiency.

This is the opposite of "business as usual" in the construction sector, where people do things on project after project in the same old inefficient ways, forcing each other to give up profits and overhead recovery in order to deliver at what seems the market price. What results is a fight over who keeps any of the meagre margins that result from each project, or attempts to recoup "negative margins" through "claims". The last thing that receives time or energy in this desperate, project-by-project, gladiatorial battle for survival is consideration of how to reduce underlying costs or improve quality.'

Whilst successful firms in all other sectors see partnering as almost exclusively a supply-side process that is critical to the competitive success of the supply chain, virtually everyone in the construction industry seems to see partnering as the process that governs the interfaces between the demand-side customers and the fragmented supply-side design and construction team members.

This perception in other business sectors that partnering is primarily a supply-side function is firmly locked into their perception of what enables a firm to compete successfully and profitably. As both *Building Down Barriers Handbook of Supply Chain Management* and Sigma Management Development Ltd's book *Supply Chain Relationships in Aerospace – Working Together* correctly emphasised, competitive success in all other sectors comes from the ability of the end supplier to use strategic supply chain partnering to drive out the unnecessary costs that are generated by the inefficient use of labour and materials and to convert them into higher profits and lower prices, whilst at the same time maintaining or driving up quality.

In all other sectors, the critical importance of measuring and eliminating unnecessary costs is well understood, as is understanding how best to do it. The exception is the construction sector, where I have yet to hear the concept of unnecessary costs even mentioned! Because the UK construction sector has yet to develop a common understanding of unnecessary costs, because virtually no firms within the industry have an improvement programme that is based on the measurement and elimination of unnecessary costs, because the construction industry rarely practises (or even understands) true supply chain management, it is not surprising that the construction industry sees little value in long-term, strategic supply chain partnerships.

This lack of understanding of the importance of unnecessary costs, coupled with the serious dearth of information about their nature and magnitude (BSRIA and CALIBRE being the only organisations that are exceptions to the rule), has caused the concept of partnering to be directed away from the relationships within the supply-side design and construction supply chain, and to be directed instead towards the relationships between demand-side customers and the individual members of the supply-side design and construction team. In fact, it is not uncommon to come across instances where this demand-side/supply-side partnering actually acts as a powerful barrier to effective supply chain integration because the demand-side customer sets up

separate partnerships with design consultancies (who develop the design in isolation from the construction team),
then, at the completion of design development, the demand-
side customer sets up a *separate* partnership with a construction contractor to construct the design produced by the
design consultancies. Not only does this prevent any possibility of the elimination of unnecessary costs, but also time
and time again the end result is a final cost that *exceeds* the
tender price by a considerable amount.

This misconception of the purpose of partnering also
completely ignores the nature of the demand-side customers, where the bulk are occasional or one-off customers who
are unable to offer the industry a continuous stream of
construction projects that would enable the design and construction teams to stay together from project to project and
thus continuously improve their performance from the
lessons they learn on each successive project.

The intermittent nature of most demand-side customers'
construction needs has caused the construction industry to
wilfully twist the purpose of partnering even further away
from what is done in other sectors. Many within the construction industry would have one-off customers of medium-
sized, short-duration projects believe that single-project partnering with a main construction contractor is all that is
required to deliver radical improvements in value for
money. In reality, the industry is subverting the true function
of partnering to mean something that will not require any
radical changes in the structure of the industry or in the way
that design and construction teams are formed for individual
projects. This restricting of partnering to the demand-side/
supply-side relationship means that the main construction
contractor will still be able to go to the market on each
project and select the firms within the project construction
supply chain on the basis of lowest cost. Frankly, this exposes a level of delusion within the construction industry that
is more than a little shameful. It also exposes a resistance to
change and an isolationist attitude that will form an un-

bridgeable barrier to any real improvement in performance or profitability, or to the delivery of radically better value to the end-user customers of buildings.

If partnering is to make any real difference to the performance of the construction industry, it must be used in precisely the same way as it has been used in other sectors. It must be firmly focused on improving the relationships between the supply-side firms that make up the design and construction supply chain. The purpose of partnering must be locked into the elimination of unnecessary costs by the supply chain firms working co-operatively together within the security of long-term, strategic supply chain partnerships.

Partnering must primarily be recognised as a supply-side tool that operates at a strategic level, that over-arches individual projects and is an essential precursor to an open and trusting culture across all the firms that need to work together within the entire design and construction supply chain. The creation of such a culture is imperative if the firms are to collaborate together to drive down unnecessary costs and drive up whole-life quality.

To summarise, if the construction industry is serious about delivering a radical improvement in performance and if it intends to import best practice in supply chain management (as every report around the world is demanding), the following definition of partnering ought to be adopted throughout the industry.

Partnering

The formation of long-term, strategic supply-side relationships between the firms to make up a design and construction supply chain that is capable of delivering a comprehensive range of building types and construction activities for a variety of demand-side customers (small and occasional as well as major repeat customers).

The primary purpose of such strategic supply-side partnering relationships is to enable the supply-side

design and construction firms to work together at both
project and strategic level to continuously drive out
unnecessary costs (caused by the inefficient utilisation
of labour and materials) and to continuously drive up
whole-life quality.

The output of such strategic supply-side partnering
should be the continuous conversion of unnecessary
costs into lower prices and higher profits, whilst improving the whole-life value of the building or the constructed product.

In the case of major repeat demand-side clients, the
nature of their long-term construction needs may warrant
the formation of alliances or partnerships with supply-side
design and construction teams because of the potential for
continuous improvement of the team's understanding of the
particular demand-side client's business needs from project
to project. However, as I explain later in Chapter 7, 'The
Client's Role in Partnering', any client in search of best value
ought to insist on hard evidence of pre-existing supply-side
partnerships between the firms that are intended to provide
the members of the supply-side design and construction
team. These pre-existing supply-side partnerships must include the key sub-contractors, trades contractors and manufacturers.

In conclusion, the overwhelming evidence from other
business sectors is that supply-side partnering is absolutely
essential to effective supply chain management, lean thinking and the delivery of best value. As a consequence,
evidence of pre-existing long-term, strategic, supply-side
partnering ought to be an essential part of any demand-
side client's selection process.

2 Why Partner At All?

To answer the question posed in the title of this chapter we need to look at how partnering operates in sectors other than the construction sector and understand why successful companies in those sectors continue to believe that supply-side partnering is fundamental to competitive success.

In other sectors, the secret of competitive success is not to pare profit margins to the bone or to force suppliers to cut their profit margins to a level where their long-term viability is damaged. Competitive success is not based on delivering goods that cost far more than the customer expects to pay, are delivered far later than originally promised, or that cost far more to maintain and operate than was allowed for in the customer's business plan. Nor is competitive success based on delivering goods that are not quite complete, do not function as efficiently as the customer requires, or contain components and materials that do not perform as well as the customer expects.

In other sectors, the secret of competitive success is based on boosting profit margins and lowering prices by the continuous reduction of unnecessary costs (that are generated by the inefficient utilisation of labour and materials). It is based on developing a deep understanding of the demand-side customer's business needs and delivering goods that satisfy (or, ideally, exceed) those needs at a price that matches that which the customer can afford to pay based on the long-term business plan. It is based on the recognition that the competitive success of the end supplier (the supplier that directly interfaces with the demand-side customer) is totally dependent on the competitive success of every firm in the supply chain. It is based on the recognition that competitive success is dependent on the supply chain operating as a virtual company where every firm in the supply chain is pulling in the same direction, where all practices and processes are aligned towards a common goal, and where a culture of openness and trust exists across all firms at all levels.

Above all, in other sectors the secret of competitive success is to work in open, collaborative and constructive partnerships with all the firms that make up the supply-side supply chain, where every firm strives to add value to every other firm's processes and no one is in the game of screwing their supplier's costs down, regardless of the effect it might have on the supplier's commercial viability.

Many in the construction industry believe the site- and craft-based nature of the industry makes it impossible to learn from other sectors. Many believe that nothing is broken in the traditional 'tried and true' processes and practices, so there is little point in trying to mend them. I believe that much of this resistance to change is caused by a lack of migration of managers between the construction sector and other sectors. Elsewhere, people and ideas regularly transfer between sectors and this has been the norm for decades. This means that companies regularly have an influx of highly intelligent and experienced managers at all levels, who are able to look at established custom and practice with fresh eyes and ask the question: 'Why do you do things this way?' More importantly, their colleagues welcome such questions and are receptive to people coming in from other sectors with fresh ideas and unblinkered eyes.

In the construction industry (including the designer consultancies) it is rare to come across anyone who has been imported from another sector, or to come across anyone who has gone out of the construction sector into another sector. Even when I do come across occasional imports from manufacturing, they are rarely listened to or used as effectively as they would be in other sectors. The construction industry continues to isolate itself and continues to believe that it is totally different from, and therefore cannot learn from, other sectors.

Until recently, this perception that the construction industry was different was reinforced at government level, since construction was dealt with by the Department of the Environment (which became the Department of Environment, Transport and the Regions – DETR) and all other sectors (except farming) were the responsibility of the Department of Trade and Industry (DTI). Now, of course, the responsibility for the construction sector (along with all the construction sector civil servants from the DETR) has been sensibly transferred to the DTI, and we should begin to see the construction sector civil servants in the DTI better understand how effective supply chain management and supply-side partnering operates in other sectors and then champion the migration of those good business practices into the construction sector. An excellent example of this transfer of best practice was shown in the UK aerospace sector in the mid 1990s. DTI officials recognised that the UK aerospace industry was in serious trouble and commissioned an organisation that had an excellent and well proven track record in facilitating performance improvement in the UK retail sector to investigate the problems and recommend the best way forward.

The resulting report advised that the UK aerospace sector's skills in effective supply chain management had been allowed to stagnate in comparison to its overseas competitors (especially in the USA) and that the most effective way forward was the development of a comprehensive training regime in supply chain management using the latest thinking and best practice from other sectors and from the most successful overseas competitors. This then needed to be deployed as quickly as possible across UK aerospace firms.

The DTI officials accepted the findings of the report, commissioned the development of the recommended training and agreed to cover the cost of training for the first year in order to kick-start the change process. The effect of this very positive and inspired cross-sectorial leadership from the DTI was a rapid improvement in the competitive performance of the UK aerospace firms.

The organisation that the DTI commissioned to report on the aerospace sector was Sigma Management Development Ltd and the training programme they developed to rectify the problems was called SCRIA (Supply Chain Relationships In Aerospace). The handbook produced by Sigma to support the training was titled *Working Together* and was about managing the inter-firm and inter-personal relationships within the supply chain in order to drive out unnecessary cost and drive up whole-life quality.

The following extracts from the Sigma handbook ought to strike a chord with those in the construction sector and should make it clear why long-term, strategic supply chain partnerships are as key to a radical improvement in performance in the construction sector as they are in other business sectors.

SCRIA – The key drivers for the aerospace industry

❑ The industry has to build products that are affordable.
❑ There is huge pressure on the industry for technological evolution, which requires huge investment with a very long payback period.
❑ To succeed, the market has to be attacked on a global scale.
❑ Customers are demanding more added value services.
❑ Operating costs must be reduced by a significant margin.
❑ Time to market is critical to become more responsive to customers and to make the industry more attractive to suppliers.

❏ Environmental legislation is forcing the requirement for quieter, more efficient aeroplanes.
❏ Safety regulations only get tougher.
❏ The industry has to manage skills and capacity shortages.

SCRIA – The legacy from the past

Much has changed over the last few years but the combined influence of the entrenched attitudes and behaviours in the industry should not be underestimated.

Prime contractor culture:
❏ Know they have the power and act like it.
❏ Don't believe in relationship building.
❏ Know best.
❏ Won't listen to suppliers.
❏ Negotiation is win–lose.
❏ Assume suppliers build in extra margin and time.
❏ Traditional.
❏ Inflexible.

Supplier culture:
❏ View relationship as war.
❏ Resentful of the power used against them.
❏ Believe they are in a no-win situation.
❏ 'We can't make them understand.'
❏ Survival of the fittest.
❏ Over-sell the expectation.
❏ Believe in the Old Boy network.
❏ Selling and servicing stops when the order is delivered.

These cultures breed:
❏ distrust
❏ secrecy
❏ frustration
❏ win–lose deals
❏ suspicion
❏ antagonism
❏ financial loss

The result is hidden costs, duplication, waste and inefficiency through the supply chain.

It seems to me that virtually everything quoted above from the SCRIA handbook applies to the construction industry. In fact, this was validated during the Building Down Barriers project because we persuaded the Amec and Laing pilot project teams to send representatives onto a standard aerospace industry SCRIA course run by Sigma. They went convinced their attendance would be a complete waste of time, but returned amazed at how relevant the course was to the construction industry!

The SCRIA course emphasises the critical importance of effective supply chain partnerships to any radical improvement in performance of the virtual company. The SCRIA handbook states:

'The enlightened companies have recognised that for the supply chain to work to its optimum, the flow of information has to be excellent. They have selected a sub-set of suppliers with whom they form closer relationships in order to facilitate the information flow. These closer relationships can be regarded as a form of partnership. The key to "oiling the wheels" of the supply chain is for companies to decide which suppliers and customers have the potential to add most value to their business and agree a form of partnership. There are no pre-defined rules for a partnership. Each one should be constructed so as to be appropriate to the relationship required.'

In the aerospace sector, all customers buy more that one aeroplane; consequently, partnerships with the end customer are always a possibility and can be beneficial to both sides because of the potential for continuous improvement. Nevertheless, the SCRIA handbook emphasises that partnering with the end customer will have limited impact on the cost or the quality of the aeroplanes. Real and substantial improvement in cost and quality can only come through long-term, strategic partnering between the firms that make up the supply-side supply chain.

In my previous book, *Performance Measurement for Construction Profitability* (see Further Reading), I gave the following case history from the retail sector that demonstrated the benefits of supply-side partnering to both the supermarket and the preferred suppliers.

Case History – Supermarket supplier

I was discussing the concept of continuous performance improvement within effective supply chain management with the Managing Director of a supplier to a major supermarket chain. He said that his firm was one of two preferred suppliers of a particular product for a major supermarket chain.

The terms of his contract with the supermarket required both suppliers to meet together with a representative of the supermarket every six months. Each supplier then had to be entirely open about any improvements they had made to both process (utilisation of labour) and product (technical innovations) over the preceding six months. They also had to be entirely open about any problems they had encountered with their performance over the preceding six months.

The purpose of these regular six-monthly sessions was to ensure that each supplier could learn from the experiences of the other supplier. Each was expected to import process or product improvements from the other and each was expected (with the active support and help of the supermarket representative) to help the other solve any outstanding process or product problems.

When I voiced concerns that giving away process or product improvements in this way could be commercially damaging to each supplier, the Managing Director said that the opposite was the case. He pointed out that in accordance with good business practice his firm had a wide customer base and only supplied some of its products to the major supermarket chain, while the remainder went to other customers. In the wider market, his firm had many competitors other than the second preferred supplier to the major supermarket chain and his firm was rarely in competition with the second preferred supplier. But because the two preferred suppliers were required to share their process and product improvements, each was more competitive in the wider market and each was able to improve their market share and their profitability when selling to that wider market.

Consequently, he was an enthusiastic supporter of the supermarket's requirement that its preferred suppliers must accept a totally open book approach to process and product improvements. As far as he could see, this approach could only become commercially risky to his firm if the two preferred suppliers dominated the entire market and were always in direct competition with each other for all their customers. Since he could see no possibility of such a duopoly situation ever arising, the supermarket's requirement that preferred suppliers share their process and product improvements was always going to be commercially beneficial to the individual preferred suppliers.

Interestingly, he said that his preferred supplier contract did have a stick as well as a carrot. If either of the two preferred suppliers refused to be open about their process or product improvements, or refused to import improvements from the other preferred supplier, their contract could be terminated. The reason for this was that the supermarket believed that continuous improvement was an essential part of their competitive success and was the main reason why they kept their position as a market leader. It was therefore essential that each preferred supplier was also an enthusiastic and committed supporter of the total continuous improvement process.

Once construction industry firms accept and understand why competitive success is more assured when it is dependent on the elimination of unnecessary costs and the delivery of the highest whole-life quality and functional performance for the lowest optimum whole-life cost, it becomes self-evident that the solution requires every firm in the design and construction supply chain to operate in an entirely new way. They must work together within the virtual supply-side company in an open and trusting culture, where each firm strives to help others within the supply chain to improve their performance by the elimination of unnecessary costs and the delivery of higher quality to the demand-side end-user customer.

Whilst partnering between the virtual supply-side company and those limited demand-side customers that can offer repeat business can be beneficial, the evidence from other sectors overwhelmingly demonstrates that real improvement in performance can only come from a radical change in culture within the supply-side of the industry.

This change of culture can only happen if the relationships in the virtual supply-side company are based on a partnering relationship where every firm (architects, engineers, quantity surveyors, construction contractors, specialist suppliers and manufacturers) believes that its competitive success and its profitability is totally dependent on the degree by which it helps others within the virtual supply-side company improve their performance, in terms of eliminating unnecessary costs caused by the inefficient utilisation of labour and materials and improving the whole life cost and performance of the constructed product (the finished building).

3 The Unchanged UK Demand for Improvement

There is a tendency for those involved with the UK construction industry to believe that the demand for radical improvement in the performance of the industry, which has grown apace since the publication of the Latham Report in 1994, is a new phenomenon. In reality, the picture is entirely the opposite, with the current demand for radical improvement merely being the latest manifestation of continuous end-user dissatisfaction that can be tracked back at least 70 years.

The only real difference between the current demand for radical improvement from end users and previous demands is that current demand is proving far more difficult to misinterpret, ignore or shrug off. The creation of powerful industry bodies, such as Rethinking Construction and the Design Build Foundation (later Be after its amalgamation with the Reading Construction Forum) has counter-balanced any attempt to dismiss the call for change. Moreover, the involvement of the Office of Government Commerce (OGC) and external public sector audit bodies, such as the National Audit Office and the Audit Commission, has accelerated and guided the pace of public sector procurement reform.

Nevertheless, it is important to see the post-*Constructing the Team* drive for radical improvement since 1994 in its full historical context in order to understand why the current demand for radical reform from end users is not just a short-term aberration. It is also important to understand the key messages that continue to be repeated in every report that has ever been published about the performance of the UK construction industry.

The first major report reviewing the performance of the UK construction industry was produced in 1929 and there have been around 13 similar reports produced between 1929 and 1994. All were inspired by client concerns about the impact on their commercial performance of the inefficiency and waste in the construction industry, and all contained remarkably similar messages.

These client concerns were effectively summed up in a book written by an architect called Alfred Bossom in 1934 entitled *Reaching for the Skies*. He went to America in the early part of the last century and became closely involved in the design and construction of skyscrapers. This taught him that construction could be treated as an engineering process in which everything is scheduled in advance and all work is carried out to an agreed timetable. The result of using these engineering techniques meant that buildings were erected more quickly than they were in the UK, yet cost no more. They yielded larger profits for both the building owner and the contractor and enabled the operatives to be paid from three to five times the wages they received in the UK.

On his return, he saw the weaknesses in the performance of the British construction industry with unblinkered eyes and became an enthusiastic advocate for radical change. In his 1934 book he states:

'*All rents and costs of production throughout Great Britain are higher than they should be because houses and factories cost too much and take too long to build. For the same reason the building industry languishes, employment in it is need-lessly precarious and some of our greatest national needs, like the clearing away of the slums and the reconditioning of our factories, are rendered almost prohibitive on the score of expense.'*

'The process of construction, instead of being an orderly and consecutive advance down the line, is all too apt to become a scramble and a muddle.'

'Bad layouts add at least 15% to the production of the cotton industry. Of how many of our steel plants and woollen mills, and even our relatively up-to-date motor works, might not the same be said? The battle of trade may easily be lost before it has fairly been opened – in the architect's designing room.'

This description of a fragmented, inefficient and adversarial industry in 1934, which damaged the commercial effectiveness of its end-user clients by being guilty of passing on unnecessarily high capital costs and poor functionality, seems little different to that described in the 1994 Latham Report or the 1998 Egan Report. In fact, the only thing that appears to be different in the 1994 and 1998 reports is the realisation that the maintenance and running costs are also unnecessarily high.

This long historical continuity of end-user client concern about the poor performance of the construction industry is well documented in a book by Mike Murray and David Langford entitled *Construction Reports 1944–98* (see Further Reading). In the conclusion, the authors state:

'The reports examined in this text have a number of recurring themes that reflect an industry inflicted with long-term illness. The content of many of the reports are strikingly similar and, indeed, the contributors in Chapter 5 commented to us that the number of similarities with the 1998 Egan Report were striking. What is evident, however, is the change in language that spans the reports. The concepts of supply chain management and lean construction are all too evident in the forerunners to Egan, but without the appropriate buzzwords. This is an important aspect of industry change. As a means to combat

an apparently volatile and unpredictable market, construction has become more reliant on advice delivered from management consultants and gurus. To some extent this relies on creating a new industry paradigm, one where management is dominant and the use of a new language indicates a commitment to radical change. Sims has argued that the most famous buzzword of all, partnering, has been hijacked by consultants and corrupted by contractors. Furthermore, many of the new ideas are repackaged common sense. This new language may indeed be the building blocks for the twenty-first century construction industry, but critics would argue that too few within the industry can "walk the talk" and even fewer can "talk the talk".'

The reason why the numerous reports between 1929 and 1994 failed to have any impact on the performance of the construction industry is because the industry continues to be blind to its failings. It is unwilling to reveal the truth to itself by measuring its performance, particularly the impact of fragmentation and adversarial attitudes on the effective utilisation of labour and materials and the lack of effective preplanning of construction activities that concerned Alfred Bossom in 1934.

This situation is made worse because clients continue to reinforce fragmentation and adversarial attitudes by insisting on using a sequential procurement process. This makes it impossible to harness the skills and knowledge of the specialist suppliers into design development because they are not brought onto the scene until after the construction contractor is appointed and the design complete. Consequently, it is impossible for them to inject buildability and 'right first time' or greater standardisation of components into the developing design.

Fortunately, the Latham Report proved a major catalyst in persuading clients to actively lead the reform movement, rather than standing to one side and expecting the industry to take the initiative. The reason for this radical change in client attitudes was that the Latham Report, for the first time, said the cost of inefficiency and waste in the industry was at least 30% of the capital cost of construction. Across the entire construction industry, this burden of unnecessary cost could amount to as much as £17 billion each year. Within the public sector annual expenditure of around £23 billion, it could amount to as much as £7 billion each year.

For individual repeat clients, the message about the high level of unnecessary cost was a powerful driver for them to take a much more active role in industry reform. The Latham Report led to the formation of powerful client groups whose sole intent was to force the pace and direction of reform. The Construction Round Table had been formed in 1992 after the demise of the National Economic Development Council (NEDO) by a small group of major repeat clients, such as BAA, McDonald's, Whitbread, Unilever and Transco. The Latham Report findings re-energised Construction Round Table members and encouraged them to take a more active and visible leadership role in the industry, which culminated in the publication of their *Agenda for Change*. The Construction Clients' Forum was formed in 1994 from a mixture of client umbrella bodies, such as the British Property Federation, and major repeat clients, such as Defence Estates. The Government Construction Clients' Panel was formed in 1997 to provide a single, collective voice for government procurement agencies and departments. In addition to these client groupings, pan-industry groups with dominant client leadership were also formed. The Reading Construction Forum was incorporated in 1995 and the Design Build Foundation was incorporated in 1997.

In 1998, the Egan Report strongly reinforced the concern of clients in the UK about the high level of inefficiency and waste and equally strongly reinforced the earlier Latham Report message of the need for integration. The Egan Report differed from earlier reports by urging the importation of best practice in supply chain management from other sectors. The report states:

'We are proposing a radical change in the way we build. We wish to see, within five years, the construction industry deliver its products to its customers in the same way as the best customer-led manufacturing and service industries.'

Figure 3.1 illustrates the evidence that supports the belief first voiced in the 1994 Latham Report that at least 30% of the capital cost of construction is consumed by unnecessary costs.

The left hand column of Figure 3.1 shows the approximate cost breakdown for a typical construction project involving a fairly traditionally constructed low-rise building with fairly traditional heating and lighting systems. Overheads and profits are around the 10% level, labour accounts for around 50% of the total costs and materials account for around 40%. These figures obviously vary, depending on

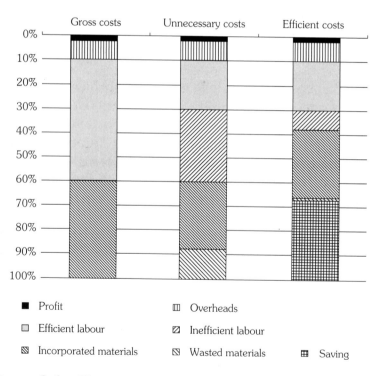

Figure 3.1   The unnecessary costs of construction

the size and complexity of the building, but in broad terms the breakdown is about right in terms of the proportions of the costs.

The centre column of Figure 3.1 takes the same overall cost breakdown but separates out the unnecessary costs in the form of the inefficient utilisation of labour and the wastage of materials. The former is caused by such things as reworking, lack of adequate preplanning, delays with the previous trades, access problems, errors in the drawings, faulty materials, insufficient or inappropriate labour, untidy or cluttered working spaces, delays in the delivery of materials, or changes to the brief. The latter is caused by such things as defective materials, incorrect sizes, scrapped materials from reworking, errors in the drawings, or changes to the brief.

The centre column of Figure 3.1 uses the BSRIA (Building Services Research and Information Association) TN 14/97 *Improving M & E Site Productivity* study and other similar studies and puts labour efficiency at the industry average of 40% of the labour element of the total costs and materials wastage at 30% of the materials element of the total costs. Obviously these are approximate figures, but the picture they represent is close to reality and shows why focusing attention on reducing unnecessary costs, rather than concentrating on the profit margin or overheads, would better improve competitiveness. Overall, the ineffective utilisation of labour and the wastage of materials put total unnecessary costs at around 42%.

The BSRIA and subsequent work done by the Construction Best Practice Programme and others has demonstrated that effective utilisation of labour is around the 30–40% level, with infrequent examples of best practice rarely exceeding 50%. The point needs to be made that this differs from the perception of senior managers, who believe that efficiency in the utilisation of labour is far higher, indeed as high as 85%. Unfortunately, this is rarely supported by accurate on-site measurement and can often be based on a perception as tenuous as 'They must be working efficiently on site because there is no one in the yard' (this was a comment by a Chief Executive of a specialist supplier firm when asked how he assessed the effective utilisation of labour in his firm).

If the submissions to the various UK construction industry awards are an indicator of the use of measurement to assess effective performance, it is rare for a submission to contain information on performance derived from measuring the effective utilisation of labour or materials. In fact, the concept of unnecessary costs seems to be alien to the construction industry, even though it has long been the main concern of firms in other sectors. Indeed, the reduction of all unnecessary costs has been the primary route by which firms in other sectors have improved their competitiveness and their profits.

In the right hand column of Figure 3.1, it has been assumed that the improved working practices that come from supply chain integration and a focus on the elimination of unnecessary costs could raise labour efficiency levels to around 70% and reduce materials wastage to around 4%. This gives a saving of around 30%, which is then available for increasing profits and wages, reducing prices, improving research and development, and developing training. Obviously these improvements are fairly conservative when compared with the performance of other sectors and if they could be bettered it ought to be possible to achieve savings of around 40–50%, as some experts on lean construction have argued. It is clear from Figure 3.1 that the insistence in the Latham Report that improved working practices could save up to 30% of the total cost of construction is well founded.

Despite the intended impact on the UK construction industry of the Latham Report in 1994 and the Egan Report in 1998, there was growing concern by the Government, by the external auditors of public sector procurement and by a small number of leading edge repeat clients, that the traditional barriers to reform were proving unassailable. It was recognised that the primary reason for this was that clients (particularly the internal professional advisors within their procurement groups) were refusing to change their traditional, sequential procurement practices and were unable to recognise that this was the main cause of the fragmentation and poor performance of the construction industry.

This led to three concomitant, but independent, moves to re-energise the reform by the publication of three best practice standards for construction procurement that could be imposed on clients by various external means. The three organisations that decided to take this proactive and courageous action were the National Audit Office, the Confederation of Construction Clients and the Cabinet Office (operating through the Department of Culture, Media and Sport, who worked with the Commission for Architecture in the Built Environment and the Treasury).

The Prime Minister became involved with the best practice standard commissioned by the Cabinet Office and he imposed the radical changes in procurement practice (set out in the report *Better Public Buildings*) on the public sector when he launched the report at 10 Downing Street in October 2000. The Confederation of Construction Clients *Clients' Charter* was launched in December 2000 and stipulates that clients can only become chartered members of the Confederation of Construction Clients if independent validation of compliance with the best practice standard set out in the *Charter Handbook* is achieved. The National Audit Office published its report *Modernising Construction* in January 2001. This comprehensive report sets out in detail the many barriers that have prevented the UK construction industry from improving its performance and describes the various initiatives that have been launched since 1994, concluding that better value means better whole-life cost and performance. It also states that better value can only be delivered by total integration of the entire design and construction team, thus ensuring that the specialist suppliers (especially the trades contractors) are involved in design from the outset. This early involvement of the trades contractors is seen by the NAO as key to the elimination of inefficiency and waste in the effective utilisation of labour and materials, the delivery of optimum whole-life costs and the delivery of maximum functionality. The NAO, not surprisingly, intends to conduct all future audits of central government procurement bodies against the best practice standard set out in *Modernising Construction*.

The importance of the impact of *Modernising Construction* and *Better Public Buildings* on the UK public sector clients should not be underestimated. The UK public sector is traditionally responsible for 40% of the total annual expenditure of the construction industry and every public sector client is subject to external audit. As a consequence, the pressure of the external auditors will force a change in the procurement practices of 40% of the industry's clients, who will be required to embrace the best practice standard that is common to *Modernising Construction* and *Better Public Buildings* in order to ensure their procurement practices deliver the best value for the tax payer.

These intense external pressures on clients will inevitably ensure that a radical reform of client procurement practice in the UK becomes irresistible. Consequently, there is an urgent need for all involved in construction procurement in the UK to understand the key requirements that are common to all the three best practice standards. Whilst this understanding should come from reading the three documents, this chapter endeavours to give busy practitioners in all sectors the priority areas for improvement.

Better Public Buildings

The report's primary thrust is targeted at the functional performance of the completed building and stipulates that well designed buildings must enhance the quality of life for the end users. In his Foreword, the Prime Minister states:

'*The best designed schools encourage children to learn. The best designed hospitals help patients recover their spirits and their health.*'

The powerful secondary thrust in the report is the need to achieve better value over the whole lifetime of the building. Again, the Prime Minister states:

'*Integrating design and construction delivers better value for money as well as better buildings, particularly when attention is paid to the full costs of a building over its whole lifetime.*'

The report demands of public sector clients radical structural and cultural change in procurement practice, with the most fundamental and far reaching being the requirement (in the 'Why and How' section) that their:

'*Procurement arrangements must enable specialist suppliers to contribute to design development from the outset.*'

This requirement should come as no surprise to anyone who has read and understood the 1998 Egan Report, which saw total integration of design and construction and the use of supply chain management as the key to better value for the end-user client.

The Better Public Buildings report concludes by listing those actions that must stop and those that must be started, the most important being as follows:

Stop
- ❑ regarding good design as an optional extra
- ❑ treating lowest cost as best value
- ❑ valuing initial capital cost as more important than whole-life cost
- ❑ imagining that effectiveness and efficiency are divorced from design

Start

- ❑ measuring efficiency and waste in construction
- ❑ appointing integrated teams focusing on the whole-life impact and performance of a development
- ❑ encouraging longer-term relationships with integrated project teams as part of long-term programmes, always subject to rigorous performance review
- ❑ using whole-life costing in the value-for-money assessment of buildings

Charter Handbook

The *Charter Handbook* closely follows the theme of *Better Public Buildings* and sets out the obligations that define a best practice client in the Rethinking Construction era. The purpose of the *Charter Handbook* is to set out and describe a best practice standard to which charter clients must commit themselves. This is illustrated in the 'Background to the Charter' section of the *Charter Handbook*, where the Construction Minister, Nick Raynsford MP, is recorded as saying that the Charter must set out:

'The minimum standards they [the clients] expect in construction procurement today, their aspirations for the future and a programme of steadily more demanding targets that will drive standards up in the future.'

The *Charter Handbook* recognises that for the current reform of the construction industry to succeed, it is imperative that the clients provide leadership for the essential changes to the structure and culture of the entire supply chain through the reform of their procurement process. The *Charter Handbook* requires chartered clients to lead the drive for continuous improvement of cultural relationships throughout the supply chain and of the constructed outputs of the industry, using performance measurement to provide proof of improvement.

The *Charter Handbook* lists the obligations of a chartered client, key of which are the following:

- ❏ Prepare a programme of cultural change with targets for its achievement, over a period of at least three years, but preferably five years or more.
- ❏ Measure their own performance in achieving their cultural change programme.
- ❏ Monitor the effects of implementing their programmes of cultural change by calculating the national Key Performance Indicators that apply to their projects.
- ❏ Review annually and amend as necessary their cultural change programme in the light of what has been achieved.

The *Charter Handbook* requires clients to have procurement processes that deliver (using measurement as the basis of proof) the following key improvements to the constructed products of the industry:

- ❏ major reductions in whole-life costs
- ❏ substantial improvements in functional efficiency
- ❏ a quality environment for end users
- ❏ reduced construction time
- ❏ improved predictability on budgets and time
- ❏ reduced defects on hand-over and during use
- ❏ elimination of inefficiency and waste in the design and construction process

The *Charter Handbook* makes clear that best practice clients should always procure buildings and constructed facilities through integrated design and construction teams (preferably tied together through long-term, strategic supply-side partnering relationships) that involve everyone in the supply chain in the design process. It also requires the client to enforce the reforms in the structural relationships, culture, process and outputs of the supply-side by making them a condition of any relationship with the construction industry. Importantly, the *Charter Handbook* also makes it clear that consultants (especially architects) must be an intrinsic part of both the industry and the integrated supply chain.

Again, these requirements should come as no surprise to anyone who has read and understood the 1994 Latham Report and the 1998 Egan Report, both of which see total integration of design and construction and the use of supply chain management as the key to better value for the end-user client.

Modernising Construction

Although this is the most detailed, comprehensive and specific of the three best practice standards, the *Modernising Construction* theme fully accords with that in *Better Public Buildings* and the *Charter Handbook*. The document is highly critical of the poor performance of the industry and the effect this has on the public purse, it states:

'In 1999, a benchmarking study of 66 central government departments' construction projects with a total value of £500 million showed that three-quarters of the projects exceeded their budgets by up to 50% and two-thirds had exceeded their original completion date by 63%.'

The document lists and describes the major barriers to improved performance of the construction industry, the key barriers being:

❑ Appointing designers separately from the rest of the team.
❑ Little integration of design teams or of the design and construction process.
❑ Insufficient weight given to user's needs and the fitness for purpose of the construction.
❑ Use by client of prescriptive specifications, which stifle innovation and restrict the scope for value for money.

❏ Design often adds to the inefficiency of the construction process.
❏ Limited use of value management.
❏ Resistance to the integration of the supply chain.
❏ Limited understanding of the true cost of construction components and processes.
❏ Limited project management skills with too strong an emphasis on crisis management.
❏ Processes are such that specialist contractors and suppliers cannot contribute their experience and knowledge to designs.

The document poses a series of key questions that public sector procurers need to consider if they are serious about improving quality and value for money. The most significant and radical of the key questions are as follows:

❏ Is supply chain integration achieved from the outset of the design process?
❏ Has the whole design and construction team been assembled before the design is well developed?
❏ What are the likely whole-life (running, maintenance and other support) costs?
❏ Have appropriate techniques been used, such as value management and value engineering, to determine whether the potential for waste and inefficiency has been minimised in the method of construction?
❏ Have efficiency improvements, to be delivered by the construction process, been quantified?

Finally, the document describes those areas where measurement of construction performance is essential, the priority areas to measure being:

❏ The cost effectiveness of the construction process, such as labour productivity on site, extent of wasted materials, and the amount of construction work that has to be redone.

❑ The quality of the completed construction and whether it is truly fit for the purpose designed.

❑ The operational efficiency of completed buildings to determine whether the improvements that the original design was intended to deliver were achieved.

Once again, these key questions should come as no surprise to anyone who has read the 1998 Egan Report, which saw total integration of design and construction and the use of supply chain management as the key to better value for the end-user client. Similarly, the priority areas where measurement is essential should not be new to anyone who has read the 1994 Latham Report, which focused on the need to integrate the design and construction team in order to eliminate the high levels of inefficiency and waste in the use of labour and materials.

It is obvious from detailed analysis of the three standards that there are two key differences and six primary goals of construction best practice that mark out best practice procurement from all other forms of traditional procurement and which are common to all three standards. These were first set out in my UK construction industry booklet *A Guide to Best Practice in Construction Procurement* and are as follows.

THE TWO KEY DIFFERENTIATORS OF CONSTRUCTION BEST PRACTICE

❑ **Abandonment of lowest capital cost as the value comparator.** This is replaced in the selection process with whole-life cost and functional performance as the value for money comparators. This means industry must predict, deliver and be measured by its ability to deliver maximum durability and functionality (which includes delighted end users).

❑ **Involving specialist contractors and suppliers in design from the outset.** This means abandoning all forms of traditional procurement that delay the appointment of the specialist suppliers (sub-contractors, specialist contractors and manufacturers) until the design is well advanced (most of the buildability problems on site are created in the first 20% of the design process). Traditional forms of sequential appointment are replaced with a requirement to appoint a totally integrated design and construction supply chain from the outset. This is only possible if the appointment of the integrated supply chain is through a single point of contact – precisely as it would be in the purchase of every other product from every other sector.

THE SIX GOALS OF CONSTRUCTION BEST PRACTICE

❑ Finished building will deliver maximum functionality, which includes delighted end users.
❑ End users will benefit from the lowest optimum cost of ownership.
❑ Inefficiency and waste in the use of labour and materials will be eliminated.
❑ Specialist suppliers will be involved in design from the outset to achieve integration and buildability.
❑ Design and construction of the building will be achieved through a single point of contact for the most effective co-ordination and clarity of responsibility.
❑ Current performance and improvement achievements will be established by measurement.

The above six goals of construction best practice were later adopted by the UK Rethinking Construction organisation and promulgated in their publication *Rethinking the Construction Client – Guidelines for Construction Clients in the Public Sector.*

Accelerating Change

In 2002, the Government stepped up the pressure for industry reform by inviting Sir John Egan to examine the performance of the industry four years after the publication of his report *Rethinking Construction*. The Government also invited him to recommend what measures should be taken to accelerate the pace of reform where it was most necessary. The approach adopted was to set up a Strategic Forum for Construction, chaired by Sir John, to seek hard evidence of change within the industry. The resulting report, *Accelerating Change*, was published in the autumn of 2002 and was intended to remind and refocus the industry on the reforms first set out in the 1998 report.

The *Accelerating Change* report demands that the industry concentrates on the delivery of high quality constructed products for the lowest optimum whole-life cost that fully satisfy the functional and economic needs of the end users over the design life of the constructed product. This requires integrated design and construction teams drawn from long-term, strategic supply-side partnerships that measure and continuously improve their performance to drive out inefficiency and waste. The report makes clear that such partnerships must be constructed so that the design can be developed with the direct involvement of specialist suppliers and manufacturers. It also states that the logistics of supplying labour and materials to site must be radically improved to enhance efficiency, reduce waste and achieve a culture of 'right first time'.

In his foreword, Sir John Egan states:

'By continuously improving its performance through the use of integrated teams, the industry will become more successful. This will in turn enable it to attract and retain the quality people it needs, which will enable it profitably to deliver products and services for its clients.'

The report sets out the key measures that are needed to accelerate the pace of change, and the foremost of these is as follows:

'By the end of 2004, 20% of construction projects by value should be undertaken by integrated teams and supply chains, and 20% of client activity by value should embrace the principles of the Clients' Charter. By the end of 2007 both these figures should rise to 50%.'

In its recommendations the report further reinforces its belief in the importance of integration and long-term strategic supply-side partnerships by stating:

'Clients should require the use of integrated teams and long-term supply chains and actively participate in their creation.'

The report emphasises the critical importance of accurate performance measurement in any continuous improvement regime by specific reference to it in its vision for the UK construction industry. Within the list of essential actions, the report calls for the following:

❑ 'A culture of continuous improvement based on performance measurement.
❑ Consistent and continuously improving performance, and improved profitability, making it highly valued by its stakeholders.'

The report endorses the value of the Rethinking Construction organisation as the main driver for accelerating the pace of change of the industry and commends its achievements since it was set up after the first Egan Report.

Rethinking the Construction Client – Guidelines for Construction Clients in the Public Sector

Interestingly, just before *Accelerating Change* was published, Rethinking Construction published its guidelines for public sector clients (which represent 40% of all clients by value). The report was entitled *Rethinking the Construction Client – Guidelines for Construction Clients in the Public Sector* and it broke new ground within the construction industry by giving a precise and unambiguous definition of construction best practice. It also used the six themes of construction best practice that formed the definition to derive six key procurement guidelines for public sector clients. The definition is as follows:

'The primary themes of construction best practice are:
❑ *The finished building will deliver maximum functionality and delight the end users.*
❑ *End users will benefit from the lowest optimum cost of ownership.*
❑ *Inefficiency and waste in the use of labour and materials will be eliminated.*
❑ *Specialist suppliers will be involved from the outset to ensure integration and buildability.*
❑ *Design and construction will be through a single point of contact.*
❑ *Performance improvement will be targeted and measurement processes put in place.'*

The six guidelines that public sector clients seeking best value should follow when procuring constructed products (such as new buildings, refurbishment, adaptations and maintenance) are given as:

❑ *'Traditional processes of selection should be radically changed because they do not lead to best value.*
❑ *An integrated team which includes the client should be formed before design and maintained throughout delivery.*
❑ *Contracts should lead to mutual benefit for all parties and be based on a target and whole-life cost approach.*
❑ *Suppliers should be selected by best value and not by lowest price: this can be achieved within EC and central government procurement guidelines.*

❑ *Performance measurement should be used to underpin continuous improvement within a collaborative working process.*

❑ *Culture and processes should be changed so that collaborative rather than confrontational working is achieved.'*

Whilst these guidelines are specifically directed at public sector clients, it is clear that the key measures of the *Accelerating Change* report apply equally well to private sector clients who are seeking best value in the constructed products they procure. The *Rethinking the Construction Client* report makes the excellent point that:

'If you don't measure, you will never know how much improvement is possible or desirable. High level targets are the starting point and are necessary to start improvement – but need to be broken down into a series of lower level targets that will enable everyone involved in a project to see how their daily work should be improved to contribute to the overall improvement target. Measures are therefore required in all these areas of activity of the critical processes so that improvement targets can be set and progress to their achievement monitored.

❑ *The determination to continuously improve overall performance throughout the organisation in a structured way, with everyone wanting to be able to do tomorrow's work better than today's, is also new to many organisations.*

❑ *Ongoing measurement and evaluation of suppliers is essential to maintaining pressure for improvement. Contractors should only retain their framework status if they achieve continuous improvement. On early projects the present levels of performance and the approaches to managing costs collaboratively will be established. On subsequent schemes continuous improvement procedures will be introduced and the resulting performance improvements monitored and documented.'*

In the 1998 Egan Report, and in every subsequent publication, accurate performance measurement throughout the supply chain is seen as a critical success factor. Performance measurement provides the evidence to show how effectively everyone in the supply chain is performing at present, and also how their performance is improving on a project-by-project or year-by-year basis.

Performance measurement is critical to best practice procurement because it provides hard evidence that the end-user clients can use to objectively select the integrated supply-side team that has been most effective at improving efficiency and can therefore deliver the lowest optimum whole-life cost. This is important because it creates a level playing field for those firms that have heavily invested in introducing supply chain management and lean construction techniques within a fully integrated design and construction team. End-user clients are able to separate those firms that are merely talking the talk and using the popular buzzwords from those firms that are walking the talk and are using supply-side partnering and supply chain management techniques to drive out unnecessary costs and drive up whole-life quality and performance in order to give end users far better whole-life value.

Demand-side clients should always bear in mind that evidence from performance measurement is the only way that the supply-side design and construction team can back up any claims they make about being able to offer an excellent service. After all, **if they don't know how well they are doing, how do they know they are doing well?**

Moreover, the performance that needs to be measured is the day-to-day performance in the effective utilisation of the labour and materials that make up 80% or more of the total cost of construction. It is not unreasonable for any demand-side client seeking the best possible value from the construction industry to make the following assumptions.

- ❑ If the supply-side design and construction team do not know how much disruption is regularly encountered on site, how can they know their performance is improving?
- ❑ If they do not know how much rework regularly occurs (including rework by the professional designers), how can they know their performance is improving?
- ❑ If they do not know how often materials or plant arrive late, how can they know their performance is improving?
- ❑ If they do not know how often cluttered work areas cause delay and disruption, how can they know their performance is improving?
- ❑ If they do not know how much disruption and reworking is caused by errors on the drawings, how can they know their performance is improving?
- ❑ If they do not know the amount of new materials and components that are wasted, how can they know their performance is improving?
- ❑ If they do not know the amount of defective materials and components that have to be sent back for replacement, how can they know their performance is improving?

4 The International Demand for Improvement

The demand from clients in the UK for radical improvements in the performance of the construction industry is not unique, it is a demand that is echoed by construction clients across most of the developed world. This is exemplified in several documents that have been published since the Latham Report in 1994.

The first of these was the 1995 report by the Construction and Building Sub-committee of the Committee on Civilian Industrial Technology (CCIT) in the USA. The CCIT is part of the National Science and Technology Council (NSTC), which is a cabinet-level group charged with setting federal technology policy. The second document was the *Construction 21* report, which was produced for the Singapore Ministry of Manpower in 1999. The third document was the report commissioned by the Australian government following their Building for Growth Action Agenda. It was titled the *Building and Construction Industries Supply Chain Project* and was published in 2001. The fourth document was *Achieving Excellence in Construction*, which was published in 2002 by the Canadian Construction Research Board in response to the Federal Government's Innovation Strategy.

1995 USA Construction and Building Sub-committee Report

The focus of the report is the effect the performance of the construction industry has on the performance of other sectors and on their ability to be competitive. It demands that the research and development sector of the construction industry adopt two priority thrusts: the development of technologies and practices that would deliver better constructed facilities, and the development of technologies and practices that would improve the health and safety of the construction workforce. It sets a requirement that the technologies and practices have to be developed for use by 2003 and would deliver specific improvements against the 1995 baseline performance. The improvements became the National Construction Goals and are as follows:

- 50% reduction in delivery time
- 50% reduction in operation, maintenance and energy costs
- 30% increase in productivity and comfort (of the occupants)
- 50% fewer occupant-related illnesses and injuries
- 50% less waste and pollution
- 50% more durability and flexibility
- 50% reduction in construction work illnesses and injuries

The report points out that the construction sector constitutes US$850 billion, which was about 13% of GDP in 1995, and that the quality of constructed facilities are vital to the competitiveness of all US industry. It emphasises the need for a whole-life viewpoint of construction to give realistic attention to the values and costs of constructed facilities. In support of this need, it cites the example of an office building, where the annual operating costs (including the salaries of the occupants) roughly equal the initial construction cost. This means that the primary value comes from the productivity of the occupants, which depends on the capability of the building to meet user needs throughout its useful life.

The full text of the seven National Construction Goals is as follows:

- ❑ '*50% reduction in delivery time. Reduction in the time from the decision to construct a new facility to its readiness for service is vital to industrial competitiveness and to project cost reduction. During the initial programming, design, procurement, construction and commissioning process, the need of the client for the facility is not being met. Needs evolve over time, so a facility long in delivery may be uncompetitive when it is finished, and the investments in producing the facility cannot be recouped until the facility is operational. The need for reduction in time to project completion is often stronger in the case of renovations and repairs of existing facilities because of interruption of ongoing business. Owners, users, designers and constructors are among the groups calling for technologies and practices for reducing delivery time.*

- ❑ *50% reduction in operation, maintenance and energy costs. Operation and maintenance costs over the life of the facility usually exceed its first cost and may do so on an annualised cost basis. As prices for energy, water, sewage, waste, communications, taxes, insurance, fire safety, plant services, etc., represent costs to society in terms of resource consumption, operation and maintenance costs also reflect the environmental qualities of the constructed facility. Therefore, reductions in operation and maintenance and energy costs benefit the general public as well as the owners and users of the facility.*

- ❑ *30% increase in productivity and comfort. Industry and government studies have shown that the annual salary costs of the occupants of a commercial or* institutional building are of the same order of magnitude as the capital cost of the building. Indeed, the purpose of the building is to shelter and support the activities of its occupants. Improvement of the productivity of the occupants (or for an industrial facility, improvement of the productivity of the process housed by the facility) is the most important performance characteristic for most buildings.

❑ **50% fewer occupant related illness and injuries.** Buildings are generally intended to support human activities, yet the environment and performance of buildings can contribute to illness and injuries for building users. Examples are avoidable injuries caused by fire or natural hazards, slips and falls, legionnaires' disease from airborne bacteria (often associated with sick building symptoms) and building damage or collapse from fire, earthquakes, or extreme winds. Sick building symptoms include irritation of eyes, nose and skin, headache and fatigue. If improvements in the quality of the indoor environment reduce days of work lost to sick days and impaired productivity, annual nationwide savings could reach billions of dollars. Criminal violence in buildings is a safety issue which can be addressed in part by building design. Reductions in illness and injuries will increase users' productivity, as well as reducing costs of medical care and litigation.

❑ **50% less waste and pollution.** Improving the performance of buildings provides major opportunities to reduce waste and pollution at every step of the construction process, from raw material extraction to final demolition and recycling of the shelter and its contents. Examples are reduced energy use and greenhouse gas emissions and reduced water consumption and waste water production. Waste and pollution also can be reduced in the construction process: construction waste is an estimated 20–30% of the volume of landfills.

❏ *50% more durability and flexibility. Durability de-notes the capability of the constructed facility to con-tinue (given appropriate maintenance) its initial performance over the intended service life. Flexibility denotes the capability to adapt the constructed facil-ity to changes in the users' needs. High durability and flexibility contribute to the life cycle quality of con-structed facilities, as they usually endure for many decades.*

❏ *50% reduction in construction work illnesses and injuries. A factor affecting international competitive-ness is the cost of injuries and diseases among con-struction workers. Although the construction workforce represents about 6% of the nation's work-force, it is estimated that the construction industry pays for about one-third of workers' compensation. Workers' compensation insurance premiums range from 7% to 100% of payroll in the construction in-dustry. Construction workers are 2.5 times more likely to die as a result of work-related trauma than workers in all other industry sectors (13.6 deaths per 100 000 construction workers, as compared to 5.5 deaths per 100 000 workers in all other industry sectors). Construction workers also experience a higher incidence of non-fatal injuries than workers in other industries.'*

It is interesting to note the similarity between the non-technical barriers listed in the report and those listed in the Latham and Egan reports in the UK in 1994 and 1998. The barriers listed in the USA report were as follows:

❏ lack of leadership
❏ adversarial relationships
❏ parochialism
❏ fragmentation of the industry
❏ inadequate owner involvement

- ❑ increasing scarcity of skilled labour
- ❑ liability

1999 Singapore *Construction 21* Report

As in the US example, *Construction 21* took a long, hard and unbiased look at the weaknesses in the performance of the construction industry in Singapore. It compared the industry in Singapore with those in Australia, Japan, Hong Kong, the UK and the USA, and it examined developments elsewhere, such as in the Netherlands and in Denmark. It argued that the current state of affairs could not be sustained and that the industry must align its performance and practices with the other sectors of the economy. The report compared the size of the Singapore construction industry with those of other countries and noted the similarity, in that it was 9.1% of GDP in Singapore, 6.3% in Australia, 10.4% in Japan, 9.7% in Hong Kong and 8% in the UK.

The primary cause of the Singaporean industry's performance weaknesses was seen as the segregation of design and construction, which formed a barrier to the consideration of buildability, savings in labour usage, ease of maintenance and safety at the design stage. This segregation reduced the efficiency of the industry and led to much rework and wastage downstream.

Construction 21 calls for paradigm shifts in the performance and perceptions of the Singapore construction industry, namely that it should:

- ❑ Change from being perceived as dirty, demanding and dangerous to being perceived as professional, productive and progressive.
- ❑ Become a knowledge industry that compares well with other technologically advanced industries.
- ❑ Adopt a distributed manufacturing approach where construction products can be manufactured off-site and brought together on-site for assembly.

❏ Adopt an integrated approach to design, construction and maintenance where there is close co-operation and collaboration between consultants, construction contractors and manufacturers, leading to the formation of synergic partnerships.

❏ Deliver cost competitiveness through higher productivity.

2002 Canadian Construction Research Board Report
Achieving Excellence in Construction

The report was prepared in response to the Federal Government's request to the Canadian construction industry for input into the Federal Government's Innovation Strategy. The report outlines the many challenges facing the construction industry in its pursuit of improvement, especially against competition from foreign firms that might be further along the road to improvement.

The report warns that construction-related research has been chronically under-funded, that the industry is severely fragmented and that an adversarial culture pervades its way of doing business. It also warns that the industry is too focused on seeking innovative solutions at project level, rather than a systemic industry-wide programme of improvement.

The report points out that the construction sector represents 12–15% of GDP and is worth around Canadian $120 billion annually. The construction industry comprises more than 150 000 contractors (general and trade), with about half being firms of only one person. These contractors employ around 850 000 workers, which is about 5% of the total Canadian workforce. This number is expected to reach around 980 000 by 2006.

The report warns that the Canadian construction industry has suffered from a lack of productivity gains and has actually experienced negative growth. It points out that if efficiency gains can be made in the construction sector, the productivity of the overall economy will increase because construction costs are the fundamental components of all economic activity. Dealing with construction productivity issues will enhance international competitiveness of all businesses, as was already stated in the US report.

The Canadian report states that the construction industry's profits are marginal and that it is still operating under an adversarial system that dates back 75 years (like the UK). Both efficiency and productivity could be improved by a more systemic approach. Interestingly, the report states:

'It is sometimes amazing that things actually do get built and that they last and perform as they were intended. This is attributable mainly to the personal pride of workmanship at the on-site location, rather than to the construction industry infrastructure.'

In its section on international drivers for change, the report includes extracts from an International Council for Research and Innovation in Building and Construction (CIB) report that was published in April 2002. The CIB report identifies the common elements in the drivers that have led to the creation of national initiatives. The common elements are:

❑ A recognition that construction accounts for a significant proportion of national economic activity and that the effectiveness of the sector has implications for other industries and for public services.
❑ A perception that construction, in contrast to other industry sectors, has not improved its use of labour and its overall productivity in recent decades and that consequently its outputs are becoming relatively more expensive.
❑ A view that a key factor in the allegedly poor performance of construction is the number of different parties who have responsibilities within the construction process, therefore a more integrated process would be desirable.

❑ Overall, a view that construction should, by integrating its internal processes and adopting new information and production techniques, seek to become more similar to manufacturing sectors.

As can be seen in the lists below, the demand-side concerns about the poor performance of the construction industry summarised by these four common elements align with the targets for performance improvement set out in the six primary goals of construction best practice I first listed in *A Guide to Best Practice in Construction Procurement* and reaffirmed in an earlier chapter in this book.

The CIB common elements in the international drivers for change

❑ A recognition that construction accounts for a significant proportion of national economic activity and that the effectiveness of the sector has implications for other industries and for the public services.

❑ A perception that construction, in contrast to other industry sectors, has not improved its use of labour

The six primary goals of construction best practice

❑ Finished buildings will deliver maximum functionality, which includes delighted end users.

❑ End users will benefit from the lowest optimum cost of ownership.

❑ Inefficiency and waste in the utilisation of labour and materials will be eliminated.

❑ Current performance and improvement achievements will be established by measurement.

❑ End users will benefit from the lowest optimum cost of ownership.

❑ Inefficiency and waste in the utilisation of labour

and its overall productivity as much as other sectors in recent decades and that consequently its outputs are becoming relatively more expensive.

❏ A view that a key factor in the allegedly poor performance of construction is the number of different parties who have responsibilities within the construction process, therefore a more integrated process is desirable.

❏ Overall, a view that construction should, by integrating its internal processes and adopting new information and production techniques, seek to become more similar to manufacturing sectors.

and materials will be eliminated.

❏ Current performance and improvement achievements will be established by measurement.

❏ Specialist suppliers will be involved in design from the outset to achieve integration and buildability.

❏ Design and construction of the building will be achieved through a single point of contact for the most effective co-ordination and clarity of responsibility.

❏ Specialist suppliers will be involved in design from the outset to achieve integration and buildability.

❏ Design and construction of the building will be achieved through a single point of contact for the most effective co-ordination and clarity of responsibility.

❏ Current performance and improvement achievements will be established by measurement.

2001 Australian *Building and Construction Industries Supply Chain Project*

The Australian construction industry is quite close in size (in proportion to GDP) to the UK construction industry, being 6.3% of GDP and employing over 700 000 people (over 7% of the labour force). The industry is also similar to most other developed countries in that 85% of the work is carried out by sub-contractors, which, the report points out, places construction ahead of most other industries in terms of outsourcing. Across Australia, 94% of trade-based enterprises employ fewer than five people, which highlights the complexity of the co-ordination task needed at project level to deliver increased value for the client.

The report was commissioned by the Department of Industry, Science and Resources in response to the Australian Government's Building for Growth Action Agenda because supply chain management was seen as the single most wide-ranging solution to the industry's problems of a high level of fragmentation, low productivity, cost and time overruns, conflicts and disputes, claims and time-consuming litigation. The report points out that the legacy of the high level of fragmentation is that the project delivery process is considered highly inefficient in comparison with other industry sectors.

Importantly, the report warns that 85% of the industry's problems are process-, not product-related, and that integration approaches, such as design-and-construct, design-for-construction, concurrent engineering, lean construction and business process re-engineering, have proved inadequate to cope with the increasing complexity of construction projects. The report lists the following consequences of the high level of fragmentation of the Australian construction industry:

❏ 'Inadequate capture, structuring, prioritisation and implementation of client needs.
❏ The fragmentation of design, fabrication and construction
data, with data generated at one level not being readily re-
used downstream.
❏ Development of pseudo-optimal design solutions.
❏ Lack of integration, co-ordination and collaboration between the various functional disciplines involved in the
life-cycle aspects of the project.
❏ Poor communication of design intent and rationale, which
leads to unwarranted design changes, inadequate design
specifications, unnecessary liability claims, and increase in
project time and cost.'

The report points out that although numerous studies have
shown that effective supply chain management is a key
element in reducing construction costs, the report's authors
could find no studies that defined what supply chain management is in the construction process. (Clearly the authors
had not come across the UK publication *Building Down
Barriers Handbook of Supply Chain Management*, which
was published in 2000, when they researched their report.)

The report also looks at developments in the UK construction industry and found that, so far, efforts to develop collaborative relationships have not been very successful. It
states that supply chain management has replaced partnering as the latest buzzword of the UK construction industry,
but that few major clients and contractors use supply chain
management. The report suggests that the term 'supply
network' better describes the full complexity of the links
among the interconnected supply entities and that an effective construction supply chain ought to be viewed as a dy-
namic network of interdependent organisations that can be
quickly reconfigured to satisfy the specific needs of a given
client.

It is obvious from these studies from the USA, Canada,
Australia and Singapore, particularly the CIB report, which
looks at the situation across all the developed countries, that

there is universal recognition that a radical improvement in the performance of the construction industry would directly boost the competitiveness of other sectors. The reason for this is very simple: the construction industry's poor perform-ance results in unnecessarily high and badly controlled initial costs, poorly predicted maintenance and running costs, and impaired functionality, which in turn adversely affects the efficiency (and therefore the cost) of the activities housed inside the building. All these unnecessary costs feed directly into the overheads of the business activity housed in the facility. Consequently, the prices charged to the customers of that business activity (be it a manufactured item or a service) are higher than they need to be and are therefore not as competitive as they could be.

Most firms in other business sectors have striven to im-prove their competitiveness by the elimination of the un-necessary costs in their manufacturing or retail supply chain that come from the inefficient use of labour and materials. At the same time, they have also improved the quality of their product to more effectively meet the aspirations of their customers.

The effectiveness with which other industries have achieved this means they now have the knowledge and skills to look closely at the procurement process that delivers their buildings and facilities and compare it with their own manu-facturing or retail processes. They can clearly see the waste that occurs at all stages of the design and construction process and equally can see that this inefficiency is caused by fragmentation and adversarial attitudes in the design and construction supply chain. Clients are worried by the inabil-ity of design and construction teams to accurately predict the cost of ownership of buildings and facilities, meaning that clients cannot factor these costs into their long-term business plans. They do not like the unforeseen risks of ownership, which come from unexpectedly having to re-place expensive components during the service life of the building or facility. As this component replacement cost has not been anticipated in the owners' long-term business plans, it invariably has to be paid out of profit margin, or by raising prices.

The reality is that demand-side customers of the construction industry throughout the developed world that have achieved their competitive success through effective supply chain management and lean thinking are demanding that their performance is replicated by the construction industry.

This was the primary demand from the 1998 Egan Report in the UK, which said:

'We wish to see, within the next five years, the construction industry deliver its products to its customers in the same way as the best consumer-led manufacturing and service industries.'

Compare this with a similar statement in the Singaporean *Construction 21* report:

'The construction industry must be aligned with the other sectors of the economy in its performance and practices.'

The USA report also contains a similar sentiment:

'Innovation in the US construction industry is an essential component for America's economic prosperity and well-being.'

This is echoed in the Canadian report:

'If efficiency gains can be made in the construction sector, the productivity of the overall economy will increase because construction costs are the fundamental components of all economic activity. Dealing with construction productivity issues will enhance international competitiveness of all businesses.'

The statements above can also be compared with the following from the Australian report:

'The legacy of this high level of fragmentation is that the project delivery process is considered highly inefficient in comparison with other industry sectors.'

This common nature of the concerns in the various national reports about the ramifications of the construction industry's poor performance for other business sectors is quite striking and this shows up in the strong similarity of the themes for improvement that each report proposes.

Every report in every country sees the importation of best practice in supply chain management from other business sectors as the primary key to reform. This is most explicit in the Australian report, but it is clearly stated or implied in the other reports. The reports also make very clear that the construction industry should avoid trying to re-invent the supply chain wheel since supply chain management tools, techniques and terminology have been highly developed and refined in other business sectors over the last few decades. In the case of the UK, the Building Down Barriers project did precisely what the various reports advocated and imported supply chain management tools and techniques from best practice in the manufacturing sector. These tools and techniques were then carefully refined and tested in use on two pilot projects before they were published by CIRIA in 2000 in the *Building Down Barriers Handbook of Supply Chain Management*. The availability of this comprehensive supply chain management toolkit, if it is taken up and used as intended, ought to put the UK construction industry ahead of the rest of the developed world.

Along with supply chain management, supply chain integration also appears as a key theme for the reform of the construction industry. The reports make clear that it is the total design and construction supply chain that needs to be integrated, which must include the trades contractors. This is emphasised in the Singapore report, but is equally obvious in all the other reports. It is also obvious from the reports that supply chain integration is to be achieved by importing supply chain management tools and techniques from other business sectors, since integration is a fundamental part of supply chain management.

Fundamental to effective supply chain management in other business areas is the elimination of unnecessary costs caused by the inefficient utilisation of labour and materials. This is reflected in another key theme that is common to all the reports, since they all demand the 'elimination of inefficiency and waste' and make clear that this refers to making far more effective use of labour and materials in the design and construction process, in much the same way that the most successful manufacturers do. This improvement in efficiency is also locked into the need to accurately measure the effective utilisation of labour and materials as any successful manufacturer would do, rather than making erroneous assumptions.

Interestingly, although the UK construction industry seems to be wedded to partnering as the solution to all its ills, the term rarely appears in any of the various national reports, other than in the context of the relationships in the design and construction supply chain. This is hardly surprising, since in other business sectors the concept of partnering is primarily about the nature of the long-term, strategic relationships in the supply-side supply chain.

In summary, the striking commonality in the concerns and the solutions in the various national reports on the poor performance of the construction industry makes it imperative that construction industry firms fully understand and readily accept the need for the radical reform demanded by its end-user clients. The construction industry can no longer get away with delivering constructed products to its clients that ignore the effect the high whole-life cost and poor whole-life performance of those products has on the competitive performance of the end-user client. It can no longer make the unjustified claim that it is different from all other sectors because it has to construct buildings in the open air and that therefore improvements in supply chain management cannot be imported from other sectors. The message from the demand-side customers across the developed world is very clear: the construction industry must embrace radical change in the form of the importation of best practice in supply chain management, and it must do so with urgency.

5 Partnering in the Virtual Company

Earlier chapters make clear that the primary objectives of the reform of the construction industry first mooted by the 1994 Latham report *Constructing the Team* were to drive out unnecessary costs generated by inefficiency and to drive up the whole-life quality and the whole-life performance of constructed products.

Chapter 4 makes clear that this focus on the more effective utilisation of labour and materials, lower whole-life costs and better functional performance has been echoed in every developed country. I demonstrated this by listing the four common elements in the international drivers for improvement from the 2002 CIP report. The USA *National Construction Goals* report, the Canadian *Achieving Excellence in Construction* report and the Australian *Building and Construction Industries Supply Chain Project* strongly emphasise the ramifications for other business sectors of the inefficiency and waste in the construction industry and all three demand that the construction industry import the well proven supply chain management tools and techniques that have underpinned productivity and efficiency improvements in other business sectors over the last few decades.

The magnitude of this improvement in performance is such that it cannot be done on a single project, but requires the same supply-side team to work together over a series of projects over several years to continuously improve the design and construction process from the lessons learned on each successive project. In an industry where the majority of clients are small and occasional, and the majority of projects are small in value, the industry cannot base the formation and operation of long-term, supply-side teams solely on a major client being able to supply a series of similar projects over a period of years. What is needed is for the supply-side design and construction firms to rethink the way they work together so that they are able to come together in long-term, supply-side partnerships, irrespective of their many and varied clients. This co-operative way of working requires the supply-side firms to base their relationships on long-term, strategic partnerships or alliances that are mutually supportive, trusting and open. Construction contractors must move away from massive supplier and sub-contractor databases of 20 000 or more firms. Site agents and buyers must give up the right to go to the market at will in order to secure the lowest possible price from suppliers and sub-contractors.

Virtually all the knowledge of how and why things go wrong on site time after time is locked up within the specialist suppliers (sub-contractors, trade contractors and manufacturers). If this knowledge is to be released and used to eliminate the causes of unnecessary costs, it will necessitate the creation of a totally different relationship between supply-side firms. It is very unlikely that the specialist suppliers are going to admit to construction contractors how much reworking is regularly done or how much disruption is regularly suffered unless their relationship with the construction contractor is secured by a long-term partnership or alliance that requires both sides to be honest and open about what goes repeatedly wrong on construction sites.

The Canadian report was very clear about the outstanding skill, knowledge and experience that exist at site level within the various trades' operatives. But it also made clear that the way the fragmented and adversarial industry worked together (or, more accurately, didn't work together) meant that the industry relied almost entirely on the project-level skills of individual workers rather than the industry's infrastructure.

As I explained in previous chapters, the message coming out of every report in every developed country is almost precisely the same: the construction industry will only improve its efficiency and its productivity if it imports supply chain integration and supply chain management tools and techniques from other business sectors, particularly the manufacturing sector. Fundamental to this is the adoption by the construction industry of the supply-side partnering techniques from other sectors that are an intrinsic part of effective supply chain management and are essential to the lean thinking aspect of supply chain management.

In earlier chapters I discussed the virtual supply-side company that is created by the formation of long-term, strategic supply-side partnerships or alliances. This means that a group of firms that constitute the entire supply chain for the design and construction of a typical building or constructed facility must use long-term, strategic supply chain partnerships to form themselves into a stable, mutually supportive supply-side alliance that works together and operates as a virtual company.

The concept of a 'virtual company' came out of the pioneering work on supply chain management done by the Building Down Barriers process development project, which was launched in 1997. This was intended to adapt best practice in supply chain management from manufacturing industry for use in the construction industry. The output from the project was *Building Down Barriers Handbook of Supply Chain Management* (see Further Reading).

When the Building Down Barriers team had to explain how the long-term, strategic supply chain partnerships that are the foundation of the project's approach worked, it seemed logical to describe the long-term relationship between the supply-side firms as a 'virtual' relationship, since it did not necessarily require a formal contract or sub-contract, nor did it necessitate takeovers or mergers. Others have since undertaken to develop the concept of a 'virtual firm', such as the Design Build Foundation (now known as Be), and those wishing to benefit from their development work should contact them.

This concept of a 'virtual company' is not new in other business sectors. During the latter stages of the Building Down Barriers project, I came across the work of Sigma Management Development Ltd in the UK aerospace sector. I discovered that the concept of a virtual supply-side company was well understood by those who had fully embraced the techniques of effective supply chain management in the aerospace sector. It was clear that in other areas of business, companies recognise that the route to competitive success is through being better at eliminating unnecessary costs than your competitors, which in turn enables you to deliver better value products to the demand-side customer. It is well understood that this can only be achieved with the full and enthusiastic co-operation of every firm in the supply chain and that this level of co-operation can only be achieved through long-term, strategic supply chain partnering. The Sigma training handbook for the UK aerospace sector (which was sponsored by the Department of Trade and Industry) uses the term virtual company to describe the organisation that is formed from long-term, strategic supply-side partnering relationships. Since the term virtual company is already in common use in a major business sector, it would be sensible for the construction industry to use the same language to describe the organisation formed from the long-term, strategic supply-side design and construction partnerships or alliances.

The *Building Down Barriers Handbook of Supply Chain Management* says of long-term supplier relationships:

'Long-term relationships can drive up quality and drive down both capital and through-life costs for clients. At the same

time, they can increase profitability for the supply chain. These long-term relationships are likely to be with only a small number of suppliers in each key supply category, because it is not possible to invest in the kind of relationship required with a large number of organisations.'

Although all too often the cry is heard that 'Every building is unique and different', recognising and understanding that apparently differing building types are actually very similar when broken down into components, materials and processes should help in the formation of these strategic supply-side partnerships. Steel frames are common to offices, hospitals, health centres, warehouses, multiple-occupancy living accommodation, libraries, workshops, factories and hotels. Brickwork and blockwork occur in every building type, from high-rise tower blocks to housing. Windows are common to every building type, with the only real variation being the material. Electrical services are also common to all buildings, with minor variations where there is a requirement for specialist components. Even mechanical services have a considerable commonality across all building types.

The reality of the benefits that can come from a greater commonality of components and materials across differing building types was picked up in the UK in the 1998 Egan Report. The report is highly critical of the UK construction industry's unwillingness to grasp the benefits of greater standardisation of components and materials across differing building types. It states:

'We see a useful way of dealing efficiently with the complexity of construction, which is to make greater use of standardised components. We call on clients and designers to make much greater use of standardised components and measure the benefits of greater efficiency and quality that standardisation can deliver . . . Standardisation of process and components need not result in poor aesthetics or monotonous buildings. We have seen that, both in this country and abroad, the best

architects are entirely capable of designing attractive buildings that use a high degree of standardisation.'

The Egan Report also cites examples of a lack of standardisation of components in the UK, namely:

- ❏ *'Toilet pans – there are 150 different types in the UK, but only six in the USA.*
- ❏ *Lift cars – although standard products are available, designers almost invariably wish to customise these.*
- ❏ *Doors – hundreds of combinations of size, veneer and ironmongery exist.*
- ❏ *Manhole covers – local authorities have more than 30 different specifications for standard manhole covers.'*

Case History – Building Down Barriers

Evidence of what can happen when the specialist suppliers are linked closely with designers and construction contractors within long-term, strategic supply chain relationships was clearly shown on the two buildings used to test the application of the supply chain management tools and techniques being developed by the Building Down Barriers project.

These two pilot projects achieved many outstanding improvements in performance and in outputs that came directly from the involvement of specialist suppliers (trades contractors, specialist contractors and manufacturers) in design from the outset. Not only were there outstanding improvements at project level, but also the specialist suppliers could see that if they continued to work together with the designers and construction contractors at strategy level they could continue to improve their performance on a project-by-project basis.

The steel fabrication firm on one of the pilot projects achieved major savings in the capital cost of the steel frame and a major improvement in their profit margin. In addition, they were convinced they could take 15% off the capital cost of any subsequent steel frame if the design and construction team could stay together. The specialist suppliers on both pilot projects became fully convinced of the commercial benefits that could flow directly from enabling them to work with the consultant designers at a strategic level to eliminate the recurrent causes of disruption and abortive work, so that 'right first time' on site can be achieved every time for every project.

At pilot project level, this involvement of specialist suppliers in design from the outset led to a far greater use of standard components and materials. This was not imposed by the supply chain management tools and techniques or by the architect or engineers, or by the end-user client, but came solely from the direct involvement of specialist suppliers and manufacturers at concept design stage.

Examples of the improvements measured on the two buildings that came directly from this way of working together were a 20% reduction in construction time, wastage in the materials due to rework consistently below 2%, labour efficiency (time spent overall on adding value to the building) in the region of 65–70%, no reportable accidents, no claims, an absence of commercial or contractual conflict throughout the two supply chains and a high level of morale on site.

In other business sectors, notably the UK aerospace sector, the virtual company always operates as an entity when dealing with all its customers. This works well in those sectors because the demand-side customers in other sectors understand how supply chain management operates and want the benefits that effective supply chain management can deliver, i.e. lower prices and better quality. These demand-side customers always buy their products through a single point of contact because they want and expect the design and manufacture of the product to be undertaken within a virtual supply-side company.

However, in the construction industry demand-side customers have almost no experience of dealing with a single point of contact and have traditionally commissioned design separately from construction. In fact, the situation has traditionally been even worse with the demand-side customer separately commissioning the architect, the civil and structural engineer, the mechanical and electrical engineer and the quantity surveyor, before contracting with the construction contractor.

As a consequence of these traditionally fragmented approaches to procurement by demand-side customers, those supply-side firms that form themselves into a virtual company will have to cater for two totally different types of customer. They will be able to operate as a single entity for those (initially few) enlightened demand-side customers who wish to embrace the tangible benefits of effective supply chain management and thus contract through a single point of contact. But they will also need to have the flexibility to cater for the unenlightened demand-side customers who insist on continuing with their fragmented and adversarial approach to procurement.

Thus the virtual company always works together when dealing with a client whose procurement process embraces the six primary best practice goals, since such a client would want the same single point of contact and the same efficient supply chain management that would be the norm when buying non-construction products. The client's point of contact with the virtual company should be that which makes best sense to the members of the virtual company and so may not always be the construction contractor, as would be the case in traditional procurement. But for other clients, particularly the small and occasional clients, the supply-side firms in the virtual company act as the situation dictates, either operating as single entities or in cluster groupings, but still giving the client the benefit of the improvements to the construction process they have developed within their long-term, strategic supply-side partnerships.

By working together in this mutually supportive way to drive out all forms of unnecessary cost and drive up whole-life quality and performance, each firm in the virtual company will be able to improve (and ensure) its profitability and its market share, no matter what procurement approach the client adopts. This message is well understood in other business sectors, where the competitive battle is won by being the best at stripping away unnecessary costs by effectively using labour and materials. This constant year-on-year battle to eliminate unnecessary costs is the 'lean thinking' aspect of supply chain management that has given rise to the UK construction industry's latest buzzword, but the term 'lean thinking' is not yet firmly locked into the elimination of unnecessary costs in the effective utilisation of labour and materials, or firmly linked with the primary principle of supply chain management.

The creation of a virtual company on the supply-side of the industry can be driven by a major repeat client who is determined to force the pace and direction of reform in order to achieve better value from construction procurement and thus reduce its impact on their overheads. There are several excellent examples of this in the UK with clients such as Argent, BAA, McDonald's and several other major retailers that have used their skill at managing their retail supply chains to manage their design and construction supply chains.

However, the experience of other business sectors demonstrates that this passive supply-side response to demand-side pressure from major repeat clients for supply chain integration and management is neither the only, nor the best, way of driving forward the radical reforms that supply chain integration and management demands. This is because of the risk that supply-side firms may only integrate when working for that specific client and may continue to operate in an inefficient, fragmented and adversarial way for all other clients.

The more effective way of introducing supply chain integration and management, and thus the supply-side partnering relationships that create the virtual company, is where the initiative is taken by the supply-side and it becomes the way they do their business for all their clients. Where this occurs, supply chain integration and management has a far greater chance of being introduced because the firms in the virtual company are driven by a mutual recognition of the very real commercial benefits that will accrue directly to them all from the elimination of unnecessary costs and the delivery of better quality. This supply-side initiative tends to be led in the UK by major construction contractors who have recognised that the market is becoming more intelligent and discerning and that those firms who act first to radically improve the performance of themselves and their supply chains stand the greatest chance of maintaining or improving their share of the more discerning market.

This concept of a virtual company may be more easily understood if compared to the operation of a major football team such as Manchester United or a major rugby league team such as St Helens. The skills required from those selected for a given game will vary depending on the make-up of the opposing team and the more closely the skills of the home team can match or exceed those of the opposing team, the greater the likelihood of the home team winning. This necessitates the existence of a squad of more players than are needed for an individual game, so that the manager is able to choose a different team line-up for every game.

A wise and successful manager will carefully analyse the techniques and skills of the players in the opposing team to better understand the precise requirements of that game. Such a manager will also carefully study the performance track record of the opposing team as a whole to assess its strengths and weaknesses. Having thus carefully established the likely skills requirement for the specific game, the manager will compare the precise requirement with the skills, experiences and temperaments of the individual players in the full squad (which may well be double or treble the number of players needed for an individual game) and will pick a team, plus a small number of reserves, for the game.

This analogy shows that the operation of the virtual company is akin to that of a successful sports team. The full squad of strategic supply chain partners will need to number several more than would be required for a project because it must encompass the skills and experience necessary for the variety of design and construction needs that will occur across the full range of building and civil engineering projects the virtual company might wish to embrace. From this squad, the virtual company selects the team that is appropriate for a given project. This will need to include a small team of reserves to cover situations that might arise during the design and construction process, i.e. a switch from steel frame to concrete frame, or an unexpected and unforeseeable overload of an individual firm for reasons beyond the control of the virtual company.

There is another aspect of the management of a successful football or rugby team that has great relevance to the management of a successful virtual company. Consistently high performance of the team would be impossible if the players were constantly being changed for every game, without the opportunity to train together in the relatively stable, full squad. There would be no empathy or loyalty between the players, they would have virtually no understanding of each other's skills, experience and temperament, they would never have had the opportunity to regularly train or play together as a co-ordinated and mutually supportive team. The critically important need to develop team skills, loyalty and a common goal for the team (which must apply to the full squad as much as the team for a given game) would be impossible. The manager would find the selection of the most appropriate and effective team for a given game an endless and thankless task and would probably be driven to selecting the team for a given game by sticking a pin in a list! (Or by use of lowest price tendering from specialist suppliers, in the case of the construction industry.)

The successful team requires the manager to have comprehensive and detailed knowledge of every aspect of the skills and experience of the full squad. The manager must also have total confidence in the loyalty of every member of the full squad and must have total belief in their shared understanding of mutually agreed goals. With this firm foundation for the selection process, the manager can be reasonably certain that the players selected for any game can be relied on to perform as an effective, efficient, enthusiastic and mutually supportive team.

These considerations are equally relevant to the effectiveness and success of the virtual company. All too often projects suffer because the design and construction team are cobbled together for the first time and have no expectations of ever being together in the future. Worse still, most of them will have been selected on a lowest price basis, where profit margins have been squeezed to the bone and the only way of making a decent profit may well be through claims against other team members, or against the client.

Even worse is the fact that many team members will be introduced long after the game has started, since most of the specialist suppliers (sub-contractors, trades contractors, specialist contractors or manufacturers) will not be appointed until the design is well advanced or even complete. Consequently, their skill, experience and knowledge will be ignored in the development of the design, even though harnessing it from the outset would most certainly improve the cost effectiveness and the buildability of the constructed solution and thus increase the profits of all concerned.

Any hope of creating a mutually supportive, loyal and highly skilled team in this environment is clearly an impossibility, and the resulting fragmentation and adversarial attitudes is the cause of the situation which is portrayed in the National Audit Office report *Modernising Construction*, which states:

'In 1999, a benchmarking study of 66 central government departments' construction projects with a total value of £500 million showed that three-quarters of the projects exceeded their budgets by up to 50% and two-thirds had exceeded their original completion date by 63%.'

If all of us can understand and appreciate what makes a successful team in top level sport, why can we not transfer that understanding to our own construction industry? In the case of the construction industry, 80% of the team members are drawn from the specialist suppliers' sector of the industry and, because they are not part of a virtual company, they are generally selected on a project-by-project basis by the lowest price they can tender for the individual project. Consequently, their long-term security and profitability are at risk, the rate of bankruptcy is far higher than in other industries, the entry level is dangerously low and the valuable skill and experience of the specialist suppliers is hardly ever harnessed to drive out unnecessary costs and drive up quality. How often do we as individuals join the endless and constant chorus of complaints about 'cowboy builders' or 'cowboy suppliers' in our domestic lives, but operate in a manner that enables the existence of such firms when we switch to our corporate lives?

We all need to rethink the design and construction process and replicate the experience of successful firms in other sectors. They have demonstrated that their success is founded on long-term, strategic supply chain partnerships that embody the seven principles of supply chain management described in *Building Down Barriers Handbook of Supply Chain Management*. The 1998 Egan Report makes it clear that integration of design and construction and the use of best practice in supply chain management must also be the foundation of successful virtual firms in the construction industry. The report states:

'We are proposing a radical change in the way we build. We wish to see, within five years, the construction industry deliver its products to its customers in the same way as the best customer-led manufacturing and service industries.'

The seven universal principles of supply chain management are described in depth in the *Building Down Barriers*

Handbook of Supply Chain Management. The handbook explains that there is one primary, over-arching principle of effective supply chain management and six supporting principles, these are briefly as follows.

The primary principle of effective supply chain management

☐ **Compete through superior underlying value.** Key members of the supply-side design and construction supply chain work together to improve quality and durability, and to reduce underlying unnecessary costs (the labour and materials elements of the component and process costs) while increasing profits. The reduction in unnecessary costs is primarily about ending disruption and reworking on site and the achievement of a 'right first time' culture throughout the supply-side design and construction team. Key to this principle is the close collaboration between the design professionals and the specialist suppliers (sub-contractors, trades contractors and manufacturers) that can only come through long-term, strategic supply-side partnerships. This primary principle of effective supply chain management is also known as 'lean thinking', which is the latest buzzword about performance improvement in use in the UK construction industry. Both 'lean thinking' and 'competing through superior underlying value' are about enhancing your competitive position through the elimination of all forms of unnecessary costs in the design and construction process. However, the lessons learned from the experiences of effective supply chain management in other business sectors over the last few decades have shown that 'lean thinking' cannot succeed in isolation. To be effective, lean thinking or competing through superior underlying value must have the following six supporting principles in position and all six principles must be practised by all the firms in the design and construction supply chain.

The six supporting principles of effective supply chain management:

- ❑ **Define client values.** This requires all members of the supply chain (from end users to manufacturers) to work together, using formal value management techniques, to define and record the detailed business needs of the end users that must be delivered efficiently by the built solution. This ensures that the specialist supplier's workers who are constructing the building have a detailed understanding of the end user's functional requirements.
- ❑ **Establish supplier relationships.** The products and services of the specialist suppliers (sub-contractors, trades contractors and manufacturers) account for over 80% of the total cost of construction. It is therefore essential for the entire design and construction supply chain to establish better and more collaborative ways of working together, so that the skills throughout the supply chain can be harnessed and integrated to minimise waste of labour and materials. These improved ways of working should also encourage exploitation of the latest innovations in equipment, materials and building processes.
- ❑ **Integrate project activities.** This involves breaking down the construction activities into sub-systems or clusters. These are relatively independent elements of the whole building or facility, such as groundworks, frame and envelope, mechanical and electrical services or internal finishes. Within each sub-system or cluster the design, construction, material and component suppliers work together closely to develop detailed designs, construction methods and actual prices for delivery.
- ❑ **Manage costs collaboratively.** This involves 'target costing', where suppliers work backwards from the client's functional requirements and gross maximum price (maximum affordable budget). The supply chain firms, particularly in the cluster groupings, work together

to design a product that matches the required level of quality and functionality and provides a viable level of profit for all at the agreed target price (which must be within the gross maximum price).

❑ **Develop continuous improvement.** The specialist supplier members of the design and construction supply chain (sub-contractors, trades contractors and manufacturers) openly measure and assess all aspects of their current performance, especially their effective utilisation of labour and materials. The entire design and construction supply chain then agree continuous improvement targets for each firm's design or construction performance that will deliver maximum savings in underlying process and materials costs. The ultimate goal is to eliminate all the unnecessary costs caused by inefficiency.

❑ **Mobilise and develop people.** All those involved must recognise that their staff will need to learn new ways of thinking, acting, and reacting. This involves unlearning old ways and recognising the challenges to be met and the resistance and difficulties that can be expected.

At its simplest, strategic supply chain partnering (or lean partnering) is the means by which the supply-side firms work together to drive out all forms of unnecessary cost and to drive up the quality of the constructed product. It is the foundation of every supply-side firm's ability to compete effectively for work in any situation.

In other business sectors, enlightened companies have long recognised that for the supply chain to work to its optimum, the flow of information has to be excellent. They select a sub-set of suppliers with whom they form closer relationships in order to facilitate the most efficient and effective flow of information. These closer supply-side relationships can be regarded as a form of partnership. These enlightened supply-side companies well understand that the key to oiling the wheels of the supply chain is for companies to decide which suppliers have the potential to add most value to their business and to agree a form of partnership that is mutually beneficial. Invariably, these supply-side partnerships embody the seven universal principles of supply chain management, especially the primary principle of competing through superior underlying value.

It is important that the construction industry learns from other business sectors that partnering relates primarily to supply-side relationships, since effective supply chain management (especially the lean thinking aspect of supply chain management) will only work through long-term, strategic supply-side relationships that enable the design and construction firms involved to continuously improve the way they do business. These strategic supply-side relationships create a virtual company that is able to apply the lessons learned from its continuous improvement process to offer better value to all its demand-side clients, the small and occasional clients, the one-off clients and the major repeat clients.

When we buy a manufactured product from any other sector (such as a television, a car or a ship) we do not expect to have to enter into a partnership arrangement in order to ensure value for money. In most cases (as it is in the construction industry, where the majority of clients are small and occasional) the purchase will be one-off and any form of partnership between the client and the supply-side would be of limited value in driving out unnecessary costs. There will, of course, be the occasion where a large number of identical or similar products will be required over a period of time by an individual client and this may well make a partnering arrangement sensible for a particular client in a particular instance. Nevertheless, the lesson from other sectors and the message from the Egan Report is that partnering will deliver the greatest improvements in performance where it is the basis of the long-term, strategic relationships between firms on the design and construction supply-side of the industry.

Supply chain management tools and techniques and the long-term, strategic supply-side partnering relationships that are an essential component of effective supply chain management and lean thinking necessitate a radical and profound change in the way the supply-side design and construction firms operate. These changes make it essential that their Chief Executives fully understand the precise nature of the changes in working practices that must be put in place within their own firms and within the firms with which they do business. Chief Executives must also recognise the need to measure their organisation's current performance (especially the effective utilisation of labour and materials) and that of their suppliers, so that they have a firm basis from which to start the improvement process. They must then become fervent champions of the changes in working practices, because only powerful and clear-sighted leadership from the Chief Executive can make those changes happen.

The magnitude of these changes should not be underestimated: they will affect everyone in the design and construction supply chain and they will not happen without a major change programme and the investment in carefully structured training and mentoring. Measurement of performance will be difficult at first, since it has rarely been done in the construction industry, and the results may be hard to accept, both for the organisation and for the individual concerned. This will be especially so where it relates to the effective utilisation of labour and materials and the initial measurements validate the low levels assumed by the 2002 Canadian Construction Research Board Report, the 1994 Latham Report, and confirmed by the work of the Building Research Establishment CALIBRE team and by Building Services Research and Information Association Technical Note TN 14/97 (see Further Reading).

Many people in any firm trying to embrace the totality of supply chain management and construction best practice will find the radical changes involved threatening and will endeavour to thwart them and maintain the status quo. Because of the heavy baggage of established and comfortable custom and practice that they carry, many people will find it difficult to understand the reason for the new ways of working. Because of the confusion within the construction industry over the precise meaning of many buzzwords, people will be unsure about where the industry is supposed to be going and what ought to be done differently. These barriers to improvement will require each Chief Executive to ensure that the message is expressed in simple, easy to understand terms, that it is constantly reinforced and the understanding of the recipients is checked, that the message is consistent across the industry as a whole, and that the terminology used reflects the language of other business sectors.

The lesson from those organisations that have successfully improved their performance is absolutely clear. Any drive to radically improve performance will not be successful unless it is overtly led by a Chief Executive who understands the nature and magnitude of the changes in working practices and can be seen to be determined to make those changes happen. This lesson cannot be overstated; the evidence is overwhelming.

Radical change of the magnitude needed by construction industry firms that are intent on embracing construction best practice and effective supply chain management (as defined by the six goals of construction best practice or the Rethinking Construction organisation's six themes of construction best practice) will be impossible unless the radical changes in working practice are seen by everyone in the firm to be very overtly 'owned' by the Chief Executive in person.

This will impose the greatest burden for driving forward radical improvement squarely on the shoulders of the Chief Executive. Without effective and determined top-level leadership, without a shared understanding between the Chief Executives of the firms that need to work together within the long-term, strategic supply-side partnerships of what needs to be changed and why it needs to be changed, it will be impossible for anyone below the Chief Executive to instigate and facilitate the radical supply chain management changes that are needed.

The critical importance of the Chief Executive's role was illustrated by the feedback from a series of regional workshops held by the UK Rethinking Construction organisation in the autumn of 2002. The workshops, entitled 'The National Debate', were aimed at public sector clients and their construction industry suppliers. The purpose of the workshops was to assess how well the Rethinking Construction reforms were actually progressing down at grass roots level. The workshops were also to discover what was creating barriers to progress and preventing clients from changing and improving their procurement processes. The same message came out of every workshop and can be summed up as follows.

The pace of reform is seriously hindered, and in many cases halted, because it lacks powerful and committed leadership and simple, user-friendly guidance.

This simple and powerful message from those at the sharp end of the industry makes it clear that the radical improvement process needs to be driven openly, knowledgeably and directly by the Chief Executive in person. The Chief Executive needs to explain why the changes in working practices are necessary, what needs to be done differently and who needs to do it. Unless the Chief Executive explains all this in simple, comprehensible language, confusion will reign, progress will be blocked and improvement will be minimal.

6 Setting Up Strategic Supply-Side Partnerships

Earlier chapters hopefully make clear why the supply-side of the construction industry needs to embrace long-term, strategic supply-side partnering relationships between the companies and firms that constitute the entire design and construction supply chain. They also explain what the virtual company that results from the establishment of these partnerships looks like, and why the Chief Executive's role is critical to the change process.

This chapter will explain how to go about setting up long-term, strategic supply-side partnerships and will suggest a form of protocol that could be used to cement the relationships. It is important to emphasise that partnering in the construction industry will only be effective at radically improving performance if it is used in a way that replicates the experiences of other business sectors. Thus it needs to relate primarily to the way the culture of the supply-side of the industry ought to change in order to embrace supply chain management, lean thinking, whole-life performance and whole-life costing. Absolutely fundamental to this is the need to match the recognition in other business sectors that the route to commercial success is through the elimination of unnecessary costs in the effective utilisation of labour and materials throughout the design and construction supply chain.

The previous chapters also explain and clarify why the UK construction industry needs to abandon its fragmented and adversarial culture and replace it with a culture of openness, trust and continuous improvement, which can only come from long-term, strategic supply-side partnering.

If the construction industry is to deliver the better value asked for by its demand-side customers in all developed countries, as I have shown in earlier chapters, the primary target of those supply-side companies and firms that aspire to excellence must become the reduction in the level of unnecessary underlying labour and materials costs. If the construction sector is to match the performance of other types of business, it must do as they do and attack those areas of poor performance that can offer the greatest commercial benefit.

As discussed in earlier chapters, 80% or more of the total cost of construction is made up of underlying labour and materials costs. Almost invariably, these costs are generated by the main construction contractor's specialist suppliers (sub-contractors, trades contractors and manufacturers), not by the main construction contractor's own staff and operatives. It follows from this that the main construction contractor can only be as good as the collective competence of the specialist suppliers with whom the contractor works.

If this collective competence is to be developed, it will require the introduction of a different relationship between the main construction contractor and the specialist suppliers. It will also require a different relationship between the professional design team companies and the construction team firms. For these relationships to be better than the current fragmented and adversarial relationships, they need to be based on long-term, strategic supply-side partnering that overarches individual projects.

These long-term relationships enable the lessons learned on one project to be further refined and improved on in subsequent projects without the need for a major customer supplying a sufficiently large-scale building programme to provide a constant flow of projects.

Time and time again other business sectors have proved that this supply-side-driven, continuous improvement process is the only effective way of improving the cost-effective utilisation of labour and materials. These long-term, collaborative relationships between the main construction contractor and the strategic supply-side partners are based on the following:

❑ It is only through fully harnessing the skill, knowledge and experience of the key suppliers (design consultants, specialist contractors, trades contractors and manufacturers) that the main construction contractor will be able to drive down unnecessary costs, and thus drive down the capital and whole-life costs, whilst driving up the profit margins of all the companies and firms in the supply-side design and construction supply chain.

❑ Key suppliers will only contribute properly to overall performance improvements if they have the incentive of long-term, strategic supply-side partnering.

❑ Suppliers need to make commercially sensible profit margins if they are to invest the resources necessary to drive down unnecessary costs and offer innovative solutions. Once sensible profit margins are agreed and protected, it is possible to persuade suppliers to focus on reducing the unnecessary labour and materials costs in order to give the demand-side customers a lower price, since the underlying labour and materials costs are the main determinant of the price the demand-side customer has to pay.

❑ Long-term, strategic supply-side relationships enable lessons learned on one project to be transferred to the next project, thus continually improving the competitiveness of the whole supply-side design and construction supply chain.

❑ Suppliers will only sustain their relationships with contractors for as long as they continue to drive down their own costs and offer innovative solutions.

❑ All companies and firms involved in long-term, strategic partnering will deliver training that will enable their people to understand the purpose of the new way of working, the necessary changes to working practices, and the way improved performance will be measured.

The sequence of activities that ought to be followed in order to set up these mutually beneficial partnerships is shown in Figure 6.1.

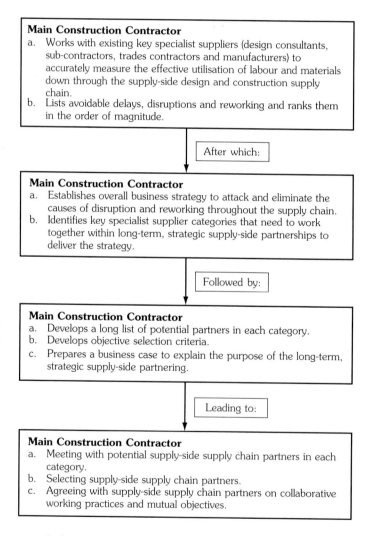

Figure 6.1 Setting up long-term strategic supply-side partnerships

ACTION PLAN FOR SETTING UP STRATEGIC SUPPLY-SIDE PARTNERSHIPS

Step 1 – Measure the effective utilisation of labour and materials

Before any thought is given to selecting specialist suppliers in a different way, the first priority is to establish (by accurate measurement) the current performance, in terms of the effective utilisation of labour and materials, of your supply chain, i.e. how well are you and your supply chain doing now?

As I have constantly emphasised in all my books, the underlying labour and materials costs of any construction project (new buildings, refurbishment, civil engineering works or maintenance) generally amounts to 80% or more of the total cost of construction. As a consequence (and as all other business sectors have long recognised) any drive for improvement must start with the accurate measurement of how effectively labour and materials are used down through the entire supply chain.

The various mechanisms for measuring the effective utilisation of labour and materials are dealt with in considerable depth in my earlier book *Performance Measurement for Construction Profitability* (see Further Reading), so I will not repeat the advice I gave in my earlier book, other than to repeat my warning that when assessing current performance, it is important to measure and not to assume, as the following case history illustrates.

Case History – Major specialist supplier

The Chief Executive of a major specialist supplier was arguing with me about the effective utilisation of labour within the construction industry and was voicing a strongly held belief that the industry generally performed very efficiently, that the complaints about poor performance were ill-founded and that he believed the performance of his own firm was indicative of the industry as a whole.

When asked how well his own firm performed, he insisted that it was beyond reproach and constantly achieved very high levels of effectiveness in its utilisation of labour. When I probed further to find out what evidence he had used to arrive at this conclusion, it transpired that he based this assumption solely on the fact that there was never any labour in the yard. He was convinced that an empty yard was unassailable proof that every one of his workers was fully and effectively utilised on site. It then became clear that he had never measured how effectively labour was utilised on site, nor did he appear to have asked his operatives how much disruption, delays or reworking they had to cope with on a regular basis. Consequently, he had no idea how much time was consumed by disruption or reworking due to errors on the drawings, defective workmanship, late deliveries of materials, congested or cluttered site conditions, poor programming of site activities, etc. Worst of all, he could see little to be gained from measuring the effective utilisation of labour on site because he could not see how he could gain commercial benefit from the expense of measuring what was actually happening on site or knowing precisely and accurately how much unnecessary cost was buried within his underlying labour and material costs.

Clearly, he had no understanding whatsoever of the well proven and firmly held belief in other business sectors that there is a direct relationship between higher profits, improved competitiveness and the elimination of unnecessary costs in the effective utilisation of labour and materials.

I should point out that his attitude is not unique, in my experience (which is supported by the experience of those involved with assessing performance under the EFQM Excellence Model), senior managers, especially CEOs and MDs, all too often fall into the trap of seeing the performance of their organisation through rose-tinted spectacles. They seldom seem to talk to the staff and operatives at the sharp end (although they often talk *at* them) and even when they do, they frequently seem unable to hear or understand what is being said to them.

CEOs and MDs have assured me many times that their organisation is achieving outstanding performance in their effective utilisation of labour and materials and that their performance levels are on a par with best practice in other business sectors. Yet when I go out and speak to their people at the sharp end, I get an entirely different picture and they complain of regularly having their work disrupted or delayed, and of having to do rework because of such things as errors on the drawings, poor preplanning of construction activities, cluttered work areas, late delivery of materials, lack of scaffolding or access problems, defective materials and clashes with other trades that are running late. This picture of disruption, delays and reworking was also confirmed when I looked at the current and historical data held by the UK Building Research Establishment CALIBRE team. In every case, assumptions about the number of visits to site by the various trades turned out to be wildly optimistic. The reality was that the number of visits regularly turned out to be increased by a factor of three or four (i.e. an assumption of three visits ended up as 11 visits) for the reasons I have listed above.

The magnitude of this gap between senior management assumptions and reality was well illustrated in the following case history from my earlier book, *Performance Measurement for Construction Profitability*.

Case History – Construction Best Practice Programme exercise

In the autumn of 2002 a *Contract Journal* article described a CBPP exercise that had taken a cross-section of construction industry firms (construction contractors and specialist suppliers) and had compared their subjective views of their effective utilisation of labour with the evidence that came from actually measuring what happened on site.

In every case, before measurements were taken, each firm insisted that it was regularly achieving labour effectiveness levels of at least 85%. Unfortunately, when the reality was carefully measured the picture was far less rosy. Not one of them exceeded 40%, with all but one coming out at between 30% and 40% in their effective utilisation of labour. The exception was measured down as low as 20%, which means the firm in question believed itself to be four times as effective in its use of labour than it really was.

If we also (not unreasonably) assume the poorest performing firm is achieving a wastage of materials due to reworking and other factors that is equally bad and is in the industry worst case band of 30% wastage, their total unnecessary cost is likely to be just over 50%. In other words, their customers are likely to be regularly paying double what they ought to be paying if the firm had achieved the effectiveness levels it claimed.

It is also important to avoid the trap of assuming all disruption and reworking occurs on site. I know from long experience that delays and reworking are equally likely to occur in the architects' or engineers' offices during the development of the design, or during the production of the working drawings, or when the selection of materials by the main construction contractor after the start on site turns out to be different to that assumed when the working drawings were completed prior to tender invitation. The following case history illustrates the situation that must occur with awful regularity in most professional offices.

Case History – Recently qualified architect

I was discussing the problem of reworking and abortive work with a young architect who had been working for a major architectural practice for a few years. He pointed out that the problem was not unique to sub-contractors. In his experience, it was not uncommon to reach site start assuming that all the working drawings were complete and that very little further work would be necessary while the project was under construction. At this stage, the project would invariably be in profit and most of the team would be redeployed to other projects.

Unfortunately, what then happened was that the construction contractor would find cheaper alternatives to the components assumed for the working drawings and all the drawings would need to be changed because the revised components had different junction details. Either that or the sub-contractors would point out that various aspects of the drawings were very difficult to construct and would therefore suggest alternative details. All of this required the architectural team to be brought back together to modify the drawings, which then delayed other projects and ate into the profit.

In the worst cases, the reworking of drawings whilst the project was on site could totally wipe out the profit that had accrued at site start and cause a great deal of antagonism between the architectural team and the construction team. It could also cause problems between the client and the architectural team.

The young architect said that if only the architectural team could work with the construction team (especially the sub-contractors or trades contractors) from the outset of the design process, there would be no need for the reworking of the drawings, there would be very little, if any, antagonism, and his firm would make far higher profits. Surely, he said, there must be a better way of working so that the sub-contractors could be available to advise the architectural team from the outset of design?

Step 2 – List the types of avoidable delay, disruption and reworking

Having worked with the existing key specialist suppliers (design consultants, specialist contractors, trades contractors and manufacturers) to measure the use of labour and materials, the next step is to work with the key specialist suppliers to list the various types of delay, disruption or reworking in order of magnitude. This enables the business strategy that is developed from the measurement exercise to attack the causes of the worst areas of unnecessary costs first and thus land an early big-win saving from the reduction in unnecessary costs that will boost morale and generate increased enthusiasm for the pace and direction of the improvement programme.

In my earlier book, *Performance Measurement for Construction Profitability*, I gave the BSRIA TN 13/2002 (see Further Reading) list of avoidable delays, disruptions and reworking as a useful guide to the kind of problems site operatives regularly face. The BSRIA TN 13/2002 list is as follows:

 (1) off-site manufacturing error
 (2) permit or method statement issue
 (3) rework through design change
 (4) inclement weather
 (5) rework through installation error
 (6) waiting for instructions
 (7) obstructed work area
 (8) spatial clash/co-ordination problem
 (9) constraint from preceding work
(10) drawing or specification issue
(11) collecting and waiting for materials
(12) collecting and waiting for plant/tools/equipment
(13) late start/early finish/extended break

Obviously items 3, 6, 8 and 10 could relate equally well to the problems encountered in an architect's or engineer's (civil and structural as well as mechanical and electrical) office, in terms of the effective utilisation of professional time. Similarly, items 2, 6, 7, 8, 9, 11, 12 and 13 could apply to the problems encountered by the main construction contractor's staff and operatives.

The outcome of Step 2 should be a very clear understanding by the main construction contractor and by all of the existing members of the main construction contractor's supply chain of the magnitude and the nature of unnecessary costs in the effective utilisation of labour and materials throughout the design and construction supply chain. It should also have engendered an understanding of the complex interrelationships that caused the unnecessary costs and an understanding of why they can only be eradicated by the firms that make up the design and construction supply chain working closely, constructively and collaboratively together, precisely as would happen in all other business sectors where supply chain management has become the norm and the elimination of unnecessary costs is the primary objective of every firm in the supply chain.

Step 3 – Establishing an overall business strategy

Having used accurate measurement to establish the effective utilisation of labour and materials down through your existing supply chain and put the types of avoidable delay, disruption and reworking in order of priority, the next thing you need to do (before you give any thought to changing the way you select and work with your specialist suppliers) is to make sure the answers to the following questions are clear in your mind:

- ❑ Why does anything need to be changed at all?
- ❑ Who will be affected by the intended changes, both within your company and externally in the form of demand-side customers and supply-side firms (design professionals, sub-contractors, trades contractors and manufacturers)?
- ❑ What aspects of performance are to be improved and how far down the supply chain will this improvement go?
- ❑ How will improved performance be measured, especially by suppliers?
- ❑ What do you anticipate will be the Critical Success Factors for the changes (the prior conditions that must be fulfilled in order to achieve an intended strategic goal)?

❑ What do you anticipate will be the Key Performance
   Results of the change (those results that it is imperative
   for the organisation to achieve)?
❑ What do you anticipate will be the Leading Indicators of
   the change? (These predict, with a degree of confidence,
   a future outcome that can be measured – such as an
   annual improvement in the effective utilisation of labour
   and materials.)

You need to ensure the strategy for your business that
evolves from the answers to these questions is in accordance
with the improvements in performance demanded by the
demand-side customers of buildings and other constructed
products. This is essential to ensure that your company does
not waste money and time pursuing an outcome that is in
conflict with the wishes of your customers.

The six goals of construction best practice I listed earlier in
this book were derived from key publications (both within
the UK and in other leading developed countries) that had
been sponsored by demand-side customers. Consequently, I
believe it would make sense for the six goals to be used as
comparative evaluators to check that your business strategy
is in line with the direction that your customers want the
construction industry to go.

The business strategy you evolve at Step 3 ought to make
the following clear to everyone that reads it:

❑ What aspects of your own company's performance and
   that of your design and construction supply chain need
   to be improved. This obviously needs to relate primarily
   to the elimination of the unnecessary costs unearthed at
   Step 1 and listed in order of priority at Step 2.
❑ How the savings generated by the improvement will
   affect the competitiveness both of your own company
   and those of your supply chain partners, i.e. how you
   intend to capture and share out the savings generated by

progressively reducing unnecessary costs on a project-by-project and year-by-year basis.

❏ How the savings generated by the improvements will affect the prices paid by demand-side customers, i.e. how much of the savings generated by progressively reducing unnecessary costs will be shared with the demand-side customers.

❏ How the improvements will affect the whole-life quality and whole-life cost of the construction works delivered to demand-side customers. This needs to assure all your current and future customers that you and your long-term, strategic supply-side partners will be driving up quality at the same time as you are driving out unnecessary costs.

❏ How you intend the supply chain companies will work differently with your company at a strategic level to achieve the necessary improvements.

❏ Over what time frame the improvements will be delivered.

❏ How those improvements will be measured down through the design and construction supply chain.

Step 4 – Identifying key specialist supplier categories

The next step is to identify the key specialist supplier categories (design professionals, sub-contractors, trades contractors and manufacturers) that need to work together and with your own company within long-term, strategic supply-side partnerships in order to eliminate the unnecessary costs you identified at Step 1, whilst at the same time improving the whole-life quality of the buildings and constructed products you deliver to the demand-side customers.

As in other business sectors, these long-term, strategic, collaborative relationships will inevitably form a virtual company, as described in earlier chapters, and as defined by Sigma Management Development Ltd in the UK Department of Trade and Industry-sponsored SCRIA handbook they developed for the aerospace sector, namely:

'The enlightened companies have recognised that for the supply chain to work to its optimum, the flow of information has to be excellent. They have selected a sub-set of suppliers with whom they form closer relationships in order to facilitate the information flow. These closer relationships can be regarded as a form of partnership.'

Many of the causes of delay, disruption and reworking on site are from errors and deficiencies in the drawings. These can only be eradicated by the constructors (the sub-contractors, trades contractors and manufacturers) working very closely at both strategic and project level with those responsible for the development of the design and the production of the working drawings (the architects, civil and structural engineers, and the mechanical and electrical engineers). Consequently, the contribution of the architectural, civil and structural engineering, and mechanical and electrical engineering companies is vital to the success of the long-term, strategic supply-side partnerships and they must figure prominently in the list of categories of key supply-side partners.

If unnecessary costs are to be eliminated, if whole-life costs are to be driven down, if whole-life quality and performance are to be driven up, both professional skills and practical expertise are equally important and equally critical to the success of the virtual company. It is imperative that every member of the supply-side design and construction team making up the virtual company values the contribution that the others in the team can make. It is especially important that the main construction contractor and the design professionals value the contribution to design development that the trades operatives can make as a result of their extensive practical experience, especially of why things do not happen as intended on site. After all, they will have been selected as strategic supply-side partners on the basis that they are experts in their own fields of activity. The following case history from *Performance*

Measurement for Construction Profitability demonstrates what can be achieved if the professional designers collaborate closely with the constructors at both project and strategic level.

Case History – Building Down Barriers

Evidence of what can happen when the specialist suppliers are linked closely with designers and construction contractors within long-term, strategic supply chain relationships was clearly shown on the two buildings used to test the application of the supply chain management tools and techniques.

These pilot projects achieved many outstanding improvements in performance and in outputs that came directly from the involvement of specialist suppliers in design from the outset. Not only were there outstanding improvements at project level, the specialist suppliers could see that if they continued to work together with the designers and construction contractors at strategy level, they could continue to improve their performance on a project-by-project basis.

The steel fabrication firm on one of the pilot projects achieved major savings in the capital cost of the steel frame and a major improvement in their profit margin. In addition, they were certain they could take 15% off the capital cost of any subsequent steel frame if the design and construction team could stay together. The specialist suppliers on both pilot projects became fully convinced of the commercial benefits that could flow directly from enabling them to work with the consultant designers at a strategic level to eliminate the recurrent causes of disruption and abortive work, so that 'right first time' on site could be achieved every time for every project.

At pilot project level, this involvement of specialist suppliers in design from the outset led to a far greater use of standard components and materials, which was not imposed by the supply chain management tools and techniques or by the architect or engineers, or by the end-user client, but came solely from the direct involvement of specialist suppliers and manufacturers at concept design stage.

Examples of the improvements measured on the two buildings that came directly from this way of working together were a 20% reduction in construction time, wastage in the materials due to rework consistently below 2%, labour efficiency (time spent overall on adding value to the building) in the region of 65–70%, no reportable accidents, no claims, an absence of commercial or contractual conflict throughout the two supply chains and a high level of morale on site.

When considering which specialist supplier categories are key to the formation of a successful virtual company, it would be sensible to look carefully at the list of types of avoidable delay, disruption and reworking drawn up in Step 2. The primary purpose of the long-term, strategic supply-side partnering is to eliminate the causes of delays, disruptions and reworking, consequently it is essential that the specialist supplier categories selected are those that need to work closely together to eliminate those causes:

❑ either because their drawings, their specifications, their preplanning of construction activities or their logistic support (in terms of the deliveries of materials and plant) appears to be causal factor in a particular area of delay, disruption or reworking;
❑ or because their operatives are particularly affected by avoidable delays, disruption and reworking and could therefore offer the most informed and constructive advice on both the causes and the cures of a specific type of delay, disruption or reworking.

For instance, if one of the items on the ranked list of avoidable delays, disruptions and reworking was the high level of non-standard components in steel frames, it would make good sense to include firms of professional structural engineers and steel fabrication firms in the list of key specialist supplier categories. In fact, it is highly likely that a similar relationship between design professionals and constructors will occur for every item on the ranked list of avoidable delays. It is also likely that this close design professional/ constructor collaboration in each category is essential if all forms of unnecessary costs are to be eliminated and both the design professionals and the site operatives are to get it right first time at every stage of a project's development.

Step 5 – Long list of potential partners

Having established the key specialist supplier categories (which, as I said, must include design team members as well as construction team members), the next step is to compose a long list of potential companies with which to establish long-term, strategic supply-side partnerships. These should ideally be drawn from those companies with which your own company has already had a long and con- structive relationship, even if the relationship has not been formally based on the ideal strategic partnering relationships described above. The list will also need to give consideration to how many firms in a given category would be needed as supply-side partners in a given geographical area for a future workload that can be deduced from your past workload.

I should warn at this point that I have often had firms bridle at my suggesting that they can safely predict their future workload based on their historic workload. They gen- erally claim that it is utterly impossible to anticipate future workloads and it is therefore impossible to forsee what supply-side specialist supplier partners they will need in a specific geographical area. Frankly, I believe this stance to be absurd because it is extremely rare for a company's work in a given geographical area to dry up overnight, and if this were true it would not be practical for them to employ any of their own staff in that area. In my experience, if a company is well established in its local area and has built up a good reputation for the quality of its work and its responsiveness to customers, the historic workload trend will continue into the future and can be safely used to decide the type and number of specialist supplier partners. In fact, if the members of the virtual company are successful at eliminating unnecessary costs and at driving up whole-life quality whilst driving down whole-life costs, their market share will increase in the geographical area in which they are operating.

Step 6 – Develop selection criteria

The process by which long-term, strategic supply-side partners are selected must be carefully structured and objective so that all those involved are convinced of its fairness and can see the range of skills, knowledge, experience and commitment that must be demonstrated in order to be selected. This is especially important for those companies that are unsuccessful: they need to know where they failed so that they know what they have to improve if they are to become part of a virtual company some time in the future.

The selection criteria will need to cover the strengths and capabilities that a specialist supplier would need to demonstrate. The criteria used to select supply-side partners will vary to some extent between professional designers and constructors, but could include the following points.

Selection criteria for strategic supply-side design partners (architects, civil and structural engineers, mechanical and electrical engineers)

❑ **Area of activity.** This has to match the type of buildings the virtual company is likely to build, or the type of construction activity it intends to undertake, and the range of demand-side customers the virtual company anticipates working with. For instance, there is little point in approaching a firm of architects whose strengths are in housing if the anticipated markets are hospitals and office blocks. It is important to carefully consider the track record of each potential design partner to ensure compatibility with the type of work the virtual company is likely to undertake.

❑ **Interest in 'buildability'.** This is a key issue because the Building Down Barriers pilot projects demonstrated that some firms of designers found it extremely difficult to accept that the trades contractors and their operatives could be equal partners and thus ought to have the right to be involved in design development from the outset, to recommend changes to the design at any stage, or that they might be capable of designing some aspects of the building themselves. All the potential design partners in the virtual company must demonstrate their willingness to respect the skill, knowledge and experience of the trades operatives and to listen to their concerns, and must also demonstrate their willingness to treat them as equals. They must demonstrate their willingness to accept that 'buildability' is an issue that is best tackled jointly by all members of the design and construction team. For instance, the architectural firm must demonstrate an enthusiastic interest in improving 'buildability' and must also demonstrate an enthusiastic willingness to work closely with the trades operatives to ensure their practical experience (particularly their knowledge of what aspects of the architect's design generally cause delay, disruption and reworking) is used to the full and incorporated into the design so that the trades operatives are able to do their work on site 'right first time'.

❑ **Getting the most cost-effective design.** The potential design partners must demonstrate an interest in developing innovative solutions or innovative approaches to the design and construction process, rather than a rigid adherence to established (and perhaps over-engineered) solutions that may be fine for professional indemnity insurance purposes but impose unjustified cost on the demand-side customer. They must demonstrate a willingness to listen to the trades operatives' suggestions and ideas, and must take them seriously. It may be worthwhile examining concerns about professional indemnity insurance at this stage, since design consultancies often cite this as the reason why they alone need to develop the design in isolation from the construction team members. Once the entire design and construction supply-side supply chain team is welded together into a virtual company, it has a very heavy impact on the traditional view of indemnity insurance. The design is no longer the sole responsibility of the design professionals, nor is there any reason why it should be when the entire design and construction team is securely locked together through supply-side partnerships. In fact, the term 'designer' in a virtual company environment will encompass the constructor partners as well as the professional design partners. In the case of the steel frame, the design will be the joint responsibility of the structural consultancy firm and the steel fabrication firm, thus, both will be jointly responsible for any long-term defects in the steel frame.

Selection criteria for strategic supply-side constructor partners (sub-contractors, trades contractors and manufacturers)

❑ **Size.** The firm needs to be large enough to mobilise the appropriate resources for the scale, complexity and type of construction activity the virtual company envisages being involved with. But it might be wise to ensure that the firm is not so large that the work they would do with the members of the virtual company would constitute such a small part of their overall business that they might not devote sufficient attention to their inter-company relationships. In particular, there might be a risk that their size might cause them to refuse to consider any form of innovation to their traditional processes or responsibilities. Such a 'take it or leave it' attitude would conflict with the collaborative partnering culture of the virtual company.

❏ **Financial status.** Has the firm the financial strength to give confidence, not only in its longer-term future, but also in its ability to make the necessary investment in training, performance measurement, the setting and monitoring of mutually agreed continuous improvement targets, and its contribution to design development from the outset of the design process? In the case of training and design involvement, these are areas that will demand considerably greater input than has previously been the case and will therefore demand a major investment of resources (people as well as money) if the firm is to be a fully committed supply-side partner in the virtual company. It might be wise for the questions exploring this criterion to demand exposure of the mechanisms (in the firm's long-term business plan) by which the necessary resources will be released from their traditional working practices and funding allocations in order to deal with these new responsibilities.

❏ **Record of innovation.** What ought to be explored here is the firm's track record in developing innovative solutions that improve the construction process, drive down unnecessary costs (in terms of the effective utilisation of labour and materials), improve durability, improve functional performance in use and reduce whole-life costs. These issues are important because, as I explained in earlier chapters, they have been picked up as key issues for demand-side customers, especially for end users, by various reports on the construction industry's performance. The new, long-term, collaborative partnerships within the virtual company will work best if every firm can demonstrate an excellent track record in coming up with innovative ideas that result in the demand-side customer receiving better value for money (preferably in whole-life cost terms).

❏ **Management style.** A successful strategic partnering relationship will only come about if the top management of each of the firms involved can see the opportunities for mutual benefit that such a relationship can deliver. They need to be willing, enthusiastic and committed partners who are able to communicate their commitment and their enthusiasm for this new way of doing business to every member of their firm. The CEO and the top management team must be seen by everyone in their firm (and by everyone that comes into contact with their firm as a customer or a supplier) to be actively and directly leading the way forward in the partnering relationships in the virtual company. The following case histories from the aerospace industry and from Defence Estates illustrate clearly the ideal management style.

Case History – Aerospace Industry

The above can be illustrated by an actual example of a major international firm from the aerospace industry that achieved a dramatic improvement in performance in a remarkably short space of time.

The Chief Executive likened his role to that of a Crusader king. He said he had to be constantly seen, by every one of his troops, to be leading the way forward into battle. Every move he made, every phrase he spoke and every word he wrote had to reinforce and clarify the changes in working practices he wanted the firm to make. He had to ensure that everyone in the firm (and he said this must literally include every last person employed in his firm) must understand where the firm was going, why it must go there, what would happen if it failed to reach its destination, and (most important) what each individual had to do differently as their part of the change process. He said it was imperative that the tea lady and the cleaners felt they were included in the change process. They must understand why the changes in working practices were commercially essential and they must want to be an active part of the change process.

He also said that it was important to recognise and reward those that were making the greatest contribution to the change process. Quite often, the reward need be no more than public recognition by the Chief Executive for their efforts (i.e. a personal letter from the Chief Executive which is also put into the firm's newsletter).

He went on to say that it was essential to provide everyone with regular progress reports which explained how the improvements in performance were being measured and what was being achieved in the various parts of the firm.

Equally important was the need to expose and deal with those who were blocking and opposing the changes in working practices. He said you could be absolutely certain that the grapevine would ensure that everyone would be aware of the names of those that were trying to block the changes. It was equally certain that everyone was watching to see if the Chief Executive was on the ball and would pick up on what was really happening. If the blockers were ignored, the message the grapevine would take around the firm was that the Chief Executive was not serious about the changes in working practices, so they could be safely ignored.

Case History – Defence Estates

The Defence Estates drive for radical improvement began with the arrival of a new Chief Executive who had been drawn from the end-user side of the UK Ministry of Defence and had already been closely involved in a successful improvement drive in another part of the Ministry. At his first Board of Directors meeting, the Directors were explaining the effectiveness of the current procurement model in terms of lowest capital cost, achievement of deadlines and the closeness of out-turn costs to tender prices (all of which compared very favourably with other major repeat clients). The new Chief Executive stopped the presentation partway through and made it plain that he viewed the current procurement model as a total disaster.

He pointed out that he was from the end-user side of the Ministry and his experience (which he insisted matched the experiences of all other end-users in the Ministry) was that the functionality of the buildings and facilities procured by

Defence Estates was generally poor, that the whole-life costs were never predicted, that the on-site construction activities always appeared to be highly inefficient with long periods of inactivity or reworking, and that the operation of buildings and facilities was all too often beset with nasty surprises from the premature failure of materials and components.

He made plain that the current approach to procurement had to be changed to give the end users what they really wanted, not what Defence Estates erroneously thought they wanted. He wanted a procurement approach that gave the end users the ability to buy buildings and facilities the way they bought all other products, i.e. a simple, one-stop approach to design and construction that enforced total integration of the design and construction process and of the design and construction supply chain, that delivered the maximum whole-life functionality for the lowest optimum whole-life cost, and that eliminated the inefficient utilisation of labour and materials (which, of course, precisely accords with the procurement approach recommended in *Rethinking Construction*).

He also made it known that those Directors who were unwilling to commit wholeheartedly to his drive for improvement should leave Defence Estates at the earliest opportunity.

❏ **Location.** How does the firm's current geographical spread of resources match the geographical area the main construction contractor sees as the area to be covered by the virtual company? Has the firm the ability to work anywhere, or will it be restricted to working close to its base? Would it be better to partner with major, national specialist suppliers, who could cover the entire intended geographical area over which the proposed virtual company will operate, or would a mix of smaller regional specialist suppliers be more appropriate? It may be that in some aspects of design and construction activity a national firm would be appropriate, whereas in other aspects a mix of smaller, regional firms would work better.

❑ **Managerial, design and technical strengths.** This needs to explore the range and depth of skilled resources available to each potential supply-side partner. Has the firm got the skilled resources to deal with new and different demands that a strategic partnering relationship based on the seven principles of supply chain management will require? Does it have people with the skill and knowledge to advise on design improvements, or to undertake a relevant aspect of the detailed design? Is the firm really expert in its field and does it have adequate technical expertise to actively and constructively contribute to technical discussions with the professional designers? Does the firm have sufficient management expertise to select and manage its own suppliers (subcontractors and manufacturers) in the same way that the main construction contractor is managing the firm itself within the virtual company? Effective supply chain management should not stop with the first tier suppliers, it should cascade all the way down through the entire supply-side supply chain. Unless a potential supply-side partner is able to demonstrate that it is able to manage effectively that part of the overall supply chain that directly supports its area of construction activity, it is unlikely ever to become a fully effective member of the virtual company.

❑ **Mutuality of interests.** Does the potential supply-side partner truly understand and believe that the supply chain management principles that will be the foundation of the long-term relationship are a better way of doing business? Is the firm able to accept that its effective utilisation of labour and materials is a good deal less that ideal and could be radically improved if it partnered with other supply-side design and construction firms in a virtual company? Is the firm willing to open its books and share the results of accurate performance measurement (especially in its effective utilisation of labour and materials) with its strategic partners? At its simplest, is the firm willing and able to become an enthusiastic crusader for the six primary goals of construction best practice?

Step 7 – Prepare the business case

Once the long list of potential partners has been drawn up and the objective selection criteria devised, the next step before meeting with the potential partners is to prepare a clear and cogent business case that explains the purpose of long-term, strategic supply-side partnering in terms that everyone can understand. This is essential if the main construction contractor's own people, and those of any potential supply-side partner, are to fully understand why this change in supply chain relationships is important to the commercial success of both the main construction contractor and of the potential supply-side partners. The business case also needs to make clear what processes will need to change to make it effective, what improved outcomes will be delivered if it is successful (in terms of greater efficiency and hence lower costs, higher profit margins and improved product performance), and how these outcomes will contribute to improved competitiveness and more satisfied demand-side customers.

It is essential that the main construction contractor is able to explain the commercial logic behind the creation of the long-term, strategic partnerships that will form the virtual company when meeting potential partners (and the firm's own people, particularly the inevitable large number of doubters) and can link that logic to their reason for approaching a potential partner. Above all, it is essential that the main construction contractor's business case ensures that potential strategic supply-side partners fully understand that the seven universal principles of supply chain management that underpin the partnerships demand intensive design effort up-front (especially on the part of the constructor members of the virtual company), demand a commitment to doing things 'right first time and every time', and demand the ability to operate within the formal disciplines of value management, value engineering, whole-life costing, continuous improvement and performance measurement. All of which require people to think in analytical and systematic ways that may differ from the crisis management approach that tends to pervade the construction industry.

Not all suppliers (design professionals, sub-contractors, trades contractors and manufacturers), when confronted with these realities set out in a clear, cogent and easy to understand business case, will want to become part of the virtual company, or, even if they want to, will have people with the requisite skills, knowledge and attitudes. It is better to find this out before entering into a long-term, strategic relationship, so the business case needs to make very clear the obligations for those involved.

The business case can be put together from the first three chapters of this book and ideally ought to be framed around the two key differentiators and the six goals of construction best practice in order to ensure the ultimate destination for the virtual company is both crystal clear and can be seen by all to be totally in line with the aspirations of demand-side customers. To remind you, the six goals of construction best practice are as follows:

- ❏ Finished building will deliver maximum functionality, which includes delighted end users.
- ❏ End users will benefit from the lowest optimum cost of ownership.
- ❏ Inefficiency and waste in the utilisation of labour and materials will be eliminated.
- ❏ Specialist suppliers will be involved in design from the outset to achieve integration and buildability.
- ❏ Design and construction of the building will be achieved through a single point of contact for the most effective co-ordination and clarity of responsibility.
- ❏ Current performance and improvement achievements will be established by measurement.

The business case could also usefully examine the major barriers to improved performance (in terms of the six goals listed above) that the long-term, strategic supply-side partnering relationships hope to eradicate. The extracts from the *Modernising Construction* report list these in Chapter 3, but the following are of particular importance to the business case:

❑ Appointing designers separately from the rest of the team.
❑ Design often adds to the inefficiency of the construction process.
❑ Resistance to the integration of the (design and construction) supply chain.
❑ Limited understanding of the true cost of construction components and processes.
❑ Limited project management skills, with a stronger emphasis on crisis management.
❑ Processes are such that specialist contractors and suppliers cannot contribute their experience and knowledge to designs.
❑ Insufficient weight given to users' needs and fitness for purpose of the construction.

The business case might also be strengthened by cross referral to some of the common elements in the international drivers for change identified in the 2002 report by the International Council for Research and Innovation in Building and Construction, namely:

❑ A perception that construction, in contrast to other industry sectors, has not improved its use of labour and its overall productivity as much as other sectors in recent decades and that consequently its outputs are becoming relatively more expensive.
❑ A view that a key factor in the allegedly poor performance of construction is the number of different

parties who have responsibilities within the construction process, thus making a more integrated process desirable.

❏ Overall, a view that construction should, by integrating its internal processes and adopting new information and production techniques, seek to become more similar to manufacturing sectors.

Step 8 – Meeting with potential partners

Using the criteria described in Step 6, 'Develop selection criteria' along with other criteria that the main construction contractor deems to be appropriate, it should be possible to reduce the long list of potential partners that was drawn up at Step 5 to a short list of firms that would be able to provide the key products and services that were identified at Step 4 as those essential to the successful formation of a long-term, strategic supply chain and from which a final choice of long-term, strategic supply-side partners could be made.

The final choice will have to be made through interviews and meetings at which both parties can ensure that they want to work together to derive mutual benefit through giving demand-side customers far better value, whilst making far better profit margins for themselves. The ultimate goal for all the supply-side firms within the virtual company is to increase their share of the main construction contractor's chosen market.

Before a main construction contractor approaches any potential strategic supply-side partner, it is essential to gather as much information about the firm as possible. This should ideally be recorded against each selection criterion so that the performance of the various short-listed firms can more easily be compared and the basis for the discussions at subsequent interviews and meetings can be more objective. This more formalised approach will also better facilitate feedback to unsuccessful firms.

The financial performance of the firms on the long list ought to be explored in depth, since an important facet of long-term, strategic supply-side partnerships is the high initial investment in expert and appropriate training, accurate performance measurement, target setting within the co-ordinated continuous improvement programmes, communication across the virtual company, and design collaboration from the outset of the design process at both strategic and project level. Unless the potential strategic supply-side partner has the financial strength to withstand this initial outlay, the firm has little chance of becoming an effective long-term partner, no matter how enthusiastic it is about this new way of doing business.

Having reduced the long list to a short list, the next move is to invite each contender to an interview to explore any areas that need further clarification. The primary purpose of such interviews is to ensure that the contender understands what the main construction contractor intends to achieve through long-term, strategic partnering relationships. This will require the main construction contractor to explain the business case developed at Step 7 in terms that can be easily understood by the short-listed firm.

It will be important to test the short-listed firm's understanding of the business case, since it is all too easy to assume an understanding that does not exist. The main construction contractor's interviewing team will clearly have a deep understanding of the business case because they developed it, but the trap (as I know from long personal experience) you can easily fall into is to assume those on the other side of the table have as much understanding of the subject as yourself because everything in the business case is so obvious! Those on the other side of the table are frequently tempted to look intelligent and knowing because they do not want to damage their chances of selection, even though they have little real understanding of what is being said to them. The following case histories illustrate this risk.

Case History – Defence Estates prime contractor initiative

On one of the early prime contracts, the Defence Estates' internal team had been interviewing the six short-listed prime contractor teams. These teams consisted of the lead firm and the supporting designers, specialist suppliers and facilities managers (as the prime contract was for the design, construction and maintenance of a series of buildings during the seven-year prime contract).

The Defence Estates' internal project manager debriefed me immediately after the interviews and said that his interviewing team had initially been impressed by all the short-listed prime contractor teams because when they sat down at the interview, each person opened their briefcase and placed a copy of my CBPP booklet *A Guide to Best Practice in Construction Procurement* on the table in front of them. The Defence Estates' team's initial assumption at seeing this gesture was to assume that those on the other side of the table would have a good understanding of the two key differentiators and the six goals of best practice listed in the booklet. These were allied very directly to what prime contracting was all about, so the Defence Estates' team naturally assumed those on the other side of the table would have a very clear understanding of the two key differentiators and the six goals of construction best practice and would be able to respond to all the questions with well founded evidence of performance improvements and cultural changes that related directly to the two key differentiators and the six goals.

Unfortunately (according to the Defence Estates' internal project manager) the reality turned out to be completely different in every case. He said that although each member of each prime contractor's team owned a CBPP booklet, their response to the questions proved beyond doubt that none of them had ever opened and read the booklet, or had any understanding of supply chain management, whole-life costing and whole-life performance on which the booklet was based.

They clearly thought that if they waved the booklet around and used the appropriate buzzwords (as no doubt was their usual practice) they would get away with it at the interviews. Needless to say, the Defence Estates' team found it very difficult to select a prime contractor's team in which they had any real confidence.

Case History – Appointment of lean construction experts

The old adage 'caveat emptor' 'buyer beware' (he alone is responsible if he is disappointed) is very apt for the selection of expert consultants.

As an illustration of the very real risk the construction industry buyer of expertise runs, during the early part of 2002 a major UK organisation was seeking experts who could advise on 'lean construction'. Several consultants who claimed expertise in this field were invited to a meeting and each was asked to define and explain the term 'lean construction'.

Unfortunately, each gave a totally different definition and neither the definition nor the associated explanation matched with the organisation's understanding of 'lean' from their knowledge of the manufacturing sector. It was therefore assumed that those attending the meeting were not the experts they claimed to be, but were exploiting a superficial knowledge of 'lean construction' to open up a new market and were hoping their superficial knowledge was greater than that of their potential client. Needless to say, such a low level of knowledge in a so-called expert would be a major barrier to improvement in the change process.

Above all, it is essential that the main construction contractor uses the interviews, and any subsequent meetings, to ensure that the potential strategic partners fully understand where the main construction contractor intends to go with the virtual company that will be formed from the strategic supply-side partnerships. The potential partners must be made to understand that their involvement with the virtual company will require intensive and continuous collaboration, at both strategy level and at project level, between design professionals, trades operatives and manufacturers in order to eradicate the causes of the ineffective utilisation of labour and materials and to deliver far better whole-life value and functionality for a far lower whole-life cost to the demand-side customer, whilst improving profit margins for every firm in the virtual company.

It is imperative that neither side of the long-term, strategic supply-side partnership enters into an agreement with any misconceptions over what the partnership will entail in terms of different ways of working. The interviews and subsequent meetings are the last opportunity to clarify both sides' understanding of what will be involved and it is therefore essential to check that there are no areas of confusion or misunderstanding left at the end of the process.

Step 9 – Selection of supply-side partners

Having decided how many strategic partners are needed in each category at Steps 4 and 5 ('Identify key specialist supplier categories' and 'Long list of potential partners'), it should be fairly easy to decide from the interviews which short-listed firms have the greatest potential to be effective partners within the virtual company.

Remember, success will only be assured if the selection process is objective with no favouritism being given to any firm because of particular personal relationships between individuals in the firm and in the main construction contractor's organisation.

The key to final selection of strategic partners is:

❑ **Understanding.** Only those firms that can demonstrate a real and deep understanding of the seven universal principles of supply chain management and the six primary goals of construction best practice should be selected as strategic supply-side partners. This should preferably be done through actual examples of changed working practices, not through assurances that might turn out not to have any basis in fact. For instance, they might be asked to show how they had been measuring their effective utilisation of labour and materials and how they then used the results to set their continuous improvement targets, or how they have been training their own people (and those of their suppliers) in supply chain management tools and techniques.

❑ **Commitment.** Only those firms that can demonstrate a total commitment (especially from the CEO and the senior management team) to implementing every aspect of supply chain management, particularly to the elimination of the unnecessary costs caused by the ineffective utilisation of labour and materials, should be selected as strategic partners. This commitment ought to be demonstrated by the firm's recent efforts at actually measuring its efficiency, or by the CEO's ability to demonstrate that he or she has a keen awareness of the types of ineffective utilisation of labour and materials from discussions with those at the sharp end in their firm.

❑ **Enthusiasm.** Only those firms whose management (especially the CEO and the senior management), staff and operatives are obviously and enthusiastically determined to collaborate with other supply-side strategic partners within the virtual company in order to drive out all forms of unnecessary cost, drive up whole-life quality, drive down whole-life costs, and drive up the profit margins of all the strategic partners should be selected. To demonstrate their determination, they ought ideally to be able to show what they are already doing to measure their performance (their effective utilisation of labour and materials) and embrace the tools and techniques of supply chain management.

The long-term, strategic supply-side partnerships will almost certainly need to be formalised by a document of some kind (not a contract, which should be a project-specific thing) that defines the relationship in terms of its objectives and the commitments that each will make to the other and to the demand-side customers. An example of such a document or protocol is suggested below. It is based on the agreement made between the two main construction contractors of the Building Down Barriers pilot projects and their specialist suppliers or cluster leaders (see Further Reading for *Building Down Barriers Handbook of Supply Chain Management*).

A PROTOCOL TO UNDERPIN THE RELATIONSHIP BETWEEN A MAIN CONSTRUCTION CONTRACTOR AND LONG-TERM, STRATEGIC, SUPPLY-SIDE PARTNERS

1. The rationale

It is our company's business objective to win work by delivering the highest whole-life quality for the lowest optimum whole-life cost in accordance with the principles of the UK Rethinking Construction movement [or of the appropriate country's improvement or innovation movement] and as defined by the following six primary goals of construction best practice:

- ❑ Finished building will deliver maximum functionality, which includes delighted end users.
- ❑ End users will benefit from the lowest optimum cost of ownership.
- ❑ Inefficiency and waste in the utilisation of labour and materials will be eliminated.
- ❑ Specialist suppliers will be involved in design from the outset to achieve integration and buildability.

❑ Design and construction of the building will be achieved through a single point of contact for the most effective co-ordination and clarity of responsibility.
❑ Current performance and improvement achievements will be established by measurement.

In our delivery of best practice to demand-side customers, we will always endeavour to reflect best practice in supply chain management and lean thinking from other business sectors. It is therefore our intention to manage our construction projects and to deal with our long-term, strategic supply-side partners in such a way that:

❑ It will be through effective preplanning and the continuous reduction of unnecessary costs (caused by the ineffective utilisation of labour and materials) that all firms in the virtual company formed by the long-term, strategic supply-side partnerships will make an appropriate and justified profit.
❑ Openness, trust, collaboration and team working will determine the relationship between us as the main construction contractor (or lead supplier) and our long-term, strategic supply-side partners.

2. The protocol

In achieving these long-term objectives we will have the responsibilities to our long-term strategic supply-side partners, and they to us, that are set out below:

❑ We will select our long-term, strategic supply-side partners on the basis of careful evaluation of their technical and managerial abilities, and by evidence of their commitment to a long-term, collaborative working relationship based on mutual benefit and on their commitment to the six goals of construction best practice.
❑ At project level, we will create cross-functional teams selected from our long-term, strategic supply-side

partners of both professional designers (architects, civil and structural engineers, mechanical and electrical engineers) and constructors (trades contractors, specialist contractors and manufacturers) for value management, value engineering and risk management so that all are involved from the outset of the design process and all are informed at all stages. Thus we will reduce to a minimum the possibility of the misunderstandings that lead to inefficiency and waste in the effective utilisation of labour and materials.

❑ We will share technical and commercial information openly (and expect the same from our partners) so that performance can be monitored and planned profits achieved.

❑ We will collaborate with our long-term, strategic supply-side partners to measure the effective utilisation of labour and materials and we will ensure the results of such measurement are made available to all in the virtual company.

❑ We will expect our long-term, strategic supply-side partners to use their measurement of performance to develop and implement continuous improvement programmes that will increase their ability to deliver better value to the demand-side customers and improved profitability to the supply-side partners in the virtual company.

❑ We will collaborate closely with our long-term, strategic supply-side partners to ensure the continuous improvement effort is used to increase our joint competitiveness and thus lead to mutual commercial advantage in a greater joint market share.

❑ We will encourage our long-term, strategic supply-side partners to involve their own supply chain firms in this new, collaborative way of working and to involve them, as appropriate, in strategic thinking or project level management of the design and construction process where their skills, experience and knowledge can be of particular value and where they have a justifiable expectation of being involved.

❑ When contracted to deliver a given project, we will undertake to pay all of our long-term, strategic supply-side partners promptly and will take steps to ensure the firms in their supply chains are treated in the same way.

Step 10 – Implementing collaborative working practices

If the long-term, strategic supply-side partnerships forming the virtual company are to produce a radically different outcome to the traditional, fragmented and adversarial ways of working, it is essential that all involved walk the talk and actually abandon their old ways of working and are seen to positively change how they do things at both strategy and project level.

In 1962, Douglas McGregor distinguished two commonly held and contrasting sets of assumptions about how people are managed and led. These were as follows:

Theory X assumptions	**Theory Y assumptions**
❑ People remain children grown larger; they naturally depend on leaders and do not want to think for themselves.	❑ People normally mature beyond childhood; they aspire to independence and responsibility, see and feel what is needed in a situation, and are capable of self-direction.
❑ People need to be told, shown and trained in proper methods of work.	❑ People who understand and care about what they are doing can devise and improve their own working methods.
❑ People need supervisors who will watch them closely and praise good work and point out errors.	❑ People need a sense that they are respected as capable of assuming responsibility and capable of correcting their own errors.

- ❏ People need specific instructions on what to do and how to do it; larger issues are none of their business.

- ❏ People want more and more understanding; they want to grasp the meaning of activities in which they are engaged; they have a desire for universal understanding and knowledge.

- ❏ People naturally resist change; they prefer things to stay 'business as usual'.

- ❏ People naturally tire of monotonous routine and enjoy new experiences; in some degree everyone is creative.

- ❏ People need to be pushed or driven.

- ❏ People need to be re-leased and encouraged and assisted.

In the construction industry, few would disagree that Theory X predominates, certainly at site level. The purpose of setting out the two models is not to imply that one is superior to the other, particularly since most leaders at times employ elements of both. However, Theory X has its limitations because it places immense responsibility on the manager or leader for dealing with problems of how one piece of work interfaces with another, since workers are not encouraged to tackle problems themselves. Construction work is typically full of interface issues and things emerging not quite as expected, so Theory X managers quickly find themselves overloaded and continually fighting fires, although reinforced in their belief that they are absolutely indispensable.

The UK National Audit Office report *Modernising Construction* warned of this tendency to crisis management and also warned that the construction industry commonly confused effective crisis management with effective project management. For the collaborative way of working that is facilitated by long-term, strategic supply-side partnerships within a virtual company that practise true supply chain management (i.e. the seven universal principles of supply chain management described in *Building Down Barriers Handbook of Supply Chain Management* – see Further Reading), making greater use of Theory Y offers a far better route to success.

Theory Y means trusting individuals and teams with much greater authority to work things out for themselves. This is made possible because the collaborative way of working together at strategy and project level ensures the full range of design and construction skills are available to work together at every opportunity. The supply chain management approach is about teams of people from across the design and construction supply chain collaborating at each stage in the design and construction process, at both the strategy level, where they can consider how best to learn from the lessons being fed back from the individual projects, and at project level, where they can apply what has been learned from previous projects and refined at strategy level. Teams such as these require the strongly participative approach to management described in Theory Y.

At project level, the supply chain management tools and techniques (which are the same as those advocated by lean construction and construction best practice) necessitate teams of people from across the design and construction supply chain working together at each stage in the overall process, first in using value-management tools and techniques to tease out precisely what value means to the demand-side customer (especially to the end users), then in devising a design strategy, developing it, and planning how to deliver it. Such mixed teams require a strongly participative approach to management.

This participative approach to management applies equally well to the strategic work on improving the performance of the partnering firms that make up the virtual company. At strategy level, the innovations and lessons learned at individual project level are picked up and communicated across all the partnering firms in the virtual company so that they can be applied on other projects for other demand-side customers. This continuous improvement process that operates at a strategic level may also involve a degree of refinement before the innovation or lesson learned is communicated across the partners in the virtual company. Such refinement is probably best done by teams of middle managers drawn from across the partnering firms and which reflect the make-up of the project-level teams. They can collaborate downwards with the individuals who are operating within their area of activity in the various project-level teams (e.g. the frame), and can also collaborate sideways with other teams of middle managers operating within other areas of project activity (e.g. the foundations and the envelope).

This cluster team approach to design and construction was first developed by the Reading Construction Forum in the UK and was explained in their publication *Unlocking Specialist Potential* (see Further Reading). The Reading approach was further developed and refined in the Building Down Barriers project and tested on the two Building Down Barriers pilot projects. An explanation of how the cluster approach works can be found in *Building Down Barriers Handbook of Supply Chain Management*.

Briefly, the cluster concept is based on the belief that radical improvement and innovation is more likely to come if the full design and construction team (that includes the specialist suppliers or trades contractors and manufacturers) is broken down into smaller autonomous teams. Each team brings together relevant professional designers, trades contractors, specialist contractors and manufacturers who focus on a relatively self-contained element of the overall building where the interfaces (physical interfaces of components and materials as well as communication interfaces between the members of the cluster team) are of critical importance if that aspect of the construction is to be built 'right first time'. Each project level cluster team needs to have very clear boundaries in terms of:

❑ The precise scope within the overall building or facility for which it is responsible. From the Building Down Barriers pilot project experience, this might be ground-works, frame and envelope, internal finishes, and mechanical and electrical services. But each virtual company will need to decide its own breakdown of construction activities.
❑ The constraints placed by the agreed design strategy.
❑ The cost target and the demand-side customer's maximum gross price for the project.

Figure 6.2 illustrates the interrelationships that cluster team working creates within a virtual company. Each cluster is a separate little supply chain for the element of the building in question and the lead firm in the cluster will have a close and direct relationship with each of the firms in the cluster supply chain. Ideally, these project relationships within the cluster team will also be long-term, strategic and geared towards the elimination of unnecessary costs and the enhancement of quality. In addition to these close and direct relationships within the cluster, the lead firm in the cluster will also have links with lead firms in other clusters to co-ordinate and preplan design and construction activities between clusters so that the construction activities of each cluster will proceed 'right first time' on site.

Obviously, Figure 6.2 is only illustrative and on a real project there would be many more cluster teams and each team would have many more member firms forming the full design and construction cluster team for the element of the building in question.

Using a participative approach to management does not mean managing in the same way under all circumstances. During the early life of the virtual company everyone is going through the learning process and will have to come to terms with new ways of working together. This 'Forming' stage in the development of the project- and strategy-level teams across the virtual company will require the manage-

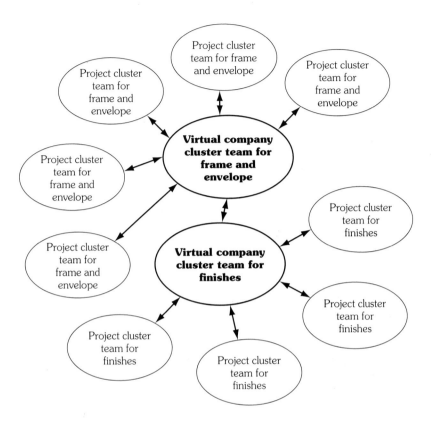

Figure 6.2 Cluster team working at project and virtual company levels

ment style to be closer to Theory X. Then, as the project and strategy level teams become more comfortable with the new ways of working together, the management style can move more to Theory Y.

A leader or manager will need to support a team in different ways at different stages of the team's development so that its members can achieve their potential for participating in the management and operation of the team. It is important to realise that team members cannot, in general, perform straight away with a high degree of self-direction, motivation and comprehension of the new ways of working. The dynamics of team formation at both project and strategy level mean that it takes a while for effective participation and its benefits to emerge, Thus participative managers will have to prepare for an initial period of carefully nurturing their team.

Table 6.1 illustrates a common model used to describe the stages of team development and the shifting forms of

Table 6.1 Leadership during team formation

Stage of development	Typical team behaviours and issues	Leadership style	Key leadership activities
Forming	Gaining an understanding of the goals and ways of working of the team, and accepting who is in it; testing out who is in charge and to what extent members can let someone else lead. For the moment, both the team's ability to do the particular task and willingness to undertake it may be low.	Telling	Making expectations clear in terms of what has already been decided, and what the group is expected to achieve in terms of the goals and ways of working. The leader needs to place greater emphasis on specifying the task and making it clear who is in overall charge.
Storming	Exploring reactions to the task and to other members of the group, including what members have in common and how they differ; expressing resistance, conflict and even hostility. The members of the team work on what concerns them about the task and the team, and begin to forge their willingness to engage.	Mobilising	Allowing expression of conflict in a managed way, so that issues stemming from differences between people and fear of each other are not suppressed and members feel free to join the team freely; in the face of challenges, explaining, rather than simply telling, the decisions concerning the task and way of working that have already been taken, and why they make sense.

Table 6.1 *Continued*

Stage of development	Typical team behaviours and issues	Leadership style	Key leadership activities
Norming	Expressing opinions and sharing information about how to proceed and co-operating in putting together an agreed plan, with an increasing level of shared responsibility and decision-making by consensus. The team as a whole develops both willingness to work and an ability to become effective at its particular task.	**Facilitating**	Sharing ideas, encouraging group decision-making in principle and in practice through offering mechanisms and formats for the team to reach democratic decisions; making sure that the team has adequate communication channels; solving problems and providing feedback when difficulties arise. The leader now places much less emphasis on telling people how to do things.
Performing	Working in an independent manner, using team effort to tackle team goals, solving problems as they arise. Team members combine working in a highly independent manner with taking account of interdependence between one another when appropriate. Both team ability and team motivation are optimal.	**Delegating**	Turning the task largely over to the members of the team, setting challenging but realistic goals; showing continued interest in how things are going and offering support when asked, but avoiding too much close supervision, which may be seen as interference; making sure that the team has adequate communication channels with other related teams.

leadership generally required to take a team through a development cycle. It suggests the key aspects of leadership
that are most important at each stage. However, most teams
will almost certainly need to revisit earlier stages of development periodically, so a leader should not slavishly follow the
suggested progression. The art of effective leadership is
often working out where a team as a whole is in terms of
its development, and then working out at that particular
moment whether the team would benefit from the application of an earlier management style. For instance, the Building Down Barriers strategy- and project-level teams were
surprisingly often pushed off course by the powerful inertia
of traditional ways of doing things. It was then necessary for
me, as the leader of the development project, to revert to a
'Telling' management style to bring the teams back on
course. In fact, the teams themselves came to recognise
their tendency to gravitate towards traditional ways of
doing things and they insisted that I regularly validated the
direction they were taking and that, if necessary, I used a
strong 'Telling' management style to pull them back on
course before they had drifted too far.

In the case of the strategic operation of the virtual company, since at the outset very few people will have a clear
understanding of what is entailed in the new working practices, it will have to go through the 'Telling', 'Mobilising',
Facilitating' and 'Delegating' stages, with the CEO of the
main construction contractor taking the initiative as the
leader of the prime firm in the virtual company of supply-
side partners. The CEO of the main construction contractor
will have done the initial 'Telling' in the business case that
was developed at Step 7 'Prepare the business case' and
issued at Step 8 'Meeting with potential suppliers'. But the
CEO needs to keep up the 'Telling' approach to the leadership of the virtual company until the CEOs of all the strategic
partners are totally on side and are 'Telling' the same message to their own people, and to the CEOs and the people in
their own supply chains. The CEO of the main construction

contractor will then have to go through the 'Mobilising', 'Facilitating' and 'Delegating' styles of management before there is any certainty that the new ways of working have totally supplanted the old outmoded customs and practices.

Exactly the same approach to management will be necessary at project level, if the new ways of working are to replace the old. There, the main construction contractor's project manager will need to take the lead in 'Telling' the integrated design and construction team, by making the expectations clear in terms of what has already been decided at strategy level in the virtual company, what the team is expected to achieve in terms of the goals, and outlining the new ways of collaborative working (especially the way that the trades contractors will be involved in design development). The project manager will then have to go through the 'Mobilising', 'Facilitating' and 'Delegating' styles of management before the team is totally comfortable with the new ways of working, but there are always likely to be times when the team will be lured off course by the power of familiar methods and they will need to be pulled back on course by the project manager using 'Telling' management style.

It is important to remember that the lower levels of management can only drive forward the new ways of working if they have the full, direct and visible support of their line managers. This is particularly true of the project-level design and construction teams where the pull of tradition will be particularly strong. Those at the lower level of management cannot be expected to drive forward new ways of working that are completely alien to the customs and practices of those in their teams unless they are totally confident that their line managers are fully behind them and will back them up if their decisions or instructions are challenged by team members from other firms.

It therefore falls to the CEO of each of the firms in the virtual company to ensure that each of the management tiers within the firm are truly walking the talk, in that they are constantly and enthusiastically reinforcing the CEO's messages about the six goals of construction best practice, supply chain management techniques, lean thinking, the elimination of unnecessary costs and better whole-life value, and are also managing and acting in ways that match their words and that facilitate the rapid introduction of the new ways of working.

The following case history from one of the Building Down Barriers pilot projects illustrates just such a need for strong and committed line management support.

Case History – Building Down Barriers pilot project

It became obvious that the young project manager of one of the two Building Down Barriers pilot projects was not providing firm leadership for his pilot project design and construction team in the new ways of working. As a result, the team was falling back into its fragmented and adversarial approach to design development, with the professional designers refusing to involve, or listen to, the specialist suppliers (sub-contractors and trades contractors).

When the problem was explored carefully and confidentially with the project manager, it became clear that although he was totally convinced of the benefits of supply chain management, total integration of design and construction, involvement of specialist suppliers from the outset of the design process, and a whole-life approach to performance and costs, his direct line managers were taking a completely opposite view and thought the whole Building Down Barriers experiment was a complete waste of time and had no hope of becoming company policy.

As a result, he was caught between a rock and a hard place. If he changed his management stance to become an openly enthusiastic and committed supporter of the Building Down Barriers approach to supply chain management, there was a very real risk that he would seriously annoy his line managers and thus kiss goodbye any chance of promotion. If, on the other hand, he adopted the antagonistic approach of his direct line managers, he would attract the fury of myself and the other senior strategic team members from the Tavistock Institute and Warwick Manufacturing Group.

He was well aware that the firm's CEO and the responsible Board Director were both committed supporters of the Building Down Barriers experiment and had earmarked a substantial contribution to its development costs. Unfortunately, he was also aware that whilst his direct line managers were publicly supportive of the CEO's commitment, they were privately against the experiment and were doing their best to sabotage it by not providing him with their support when team members challenged him over the new working practices.

Once we, as the strategic leaders of the development project, understood his predicament, we approached the Regional Manager and the Executive Board Director responsible for the firm's involvement in the Building Down Barriers experiment and explained to them what was happening and what the ramifications would be for their pilot project if nothing were done to improve the situation. They were warned that if they did not take action to provide more supportive line managers for their project manager, there was a very serious risk of their project being terminated.

They listened and accepted the need to ensure their project manager had fully supportive line managers who were committed to making the project a success. They then changed the relevant line managers and brought in people who were totally committed to the integration of the supply chain, to the use of supply chain management tools and techniques, and to the involvement of specialist suppliers from the outset of the design process (even to the extent of endorsing the project manager's proposal to change the architectural and structural engineering firms because they refused to collaborate with the specialist suppliers during design development).

This change in line managers caused a total transformation in the effectiveness of the pilot project design and construction team. The young project manager was able to show his true calibre and, under his skilful and enthusiastic management and facilitation, his pilot project rapidly caught up with, and then overtook, the other pilot project.

If the collaborative working practices are to become the way firms in the virtual company always do their business, the importance of committed and determined leadership at all levels cannot be over-emphasised. Unless *every* leader, from the CEO of the main construction contractor across to the CEOs of all the partners within the virtual company, and down through all the levels of management in each firm, are actively and enthusiastically driving their teams in the same direction and at the same speed, the intended benefits will never be fully realised.

In fact, precisely this message comes to me from those at the sharp end whenever I discuss the pace and direction of the reform of the UK construction industry with them. Their message is very clear and very simple and is summed up as follows:

'The pace of reform is seriously hindered, and in many cases halted, because it lacks powerful and committed leadership.'

This leadership requires a clear and consistent message about what has to be done differently by the supply-side of the construction industry, what the improvements will be for the demand-side customers, and how the improvements in the utilisation of labour and materials will be measured. To ensure that this message is both consistent and crystal-clear, it would make sense to lock the message very tightly into the six goals of construction best practice listed earlier in this book, namely:

❏ The finished building will deliver maximum functionality, which includes delighted end users.
❏ End users will benefit from the lowest optimum cost of ownership.
❏ Inefficiency and waste in the utilisation of labour and materials will be eliminated.
❏ Specialist suppliers will be involved in design from the outset to achieve integration and buildability.

❑ Design and construction of the building will be achieved through a single point of contact for the most effective co-ordination and clarity of responsibility.

❑ Current performance and improvement achievements will be established by measurement.

Finally, it must never be forgotten that every strategic partner in the virtual company needs to constantly check on the rate of progress and to share the result with all the other partners. This means measuring all the leading indicators of improvement on a regular basis, which means regularly measuring those things that will contribute the most to higher profit margins and lower prices.

As I have constantly emphasised in all my books, and as the various reports on the construction industry from across the developed world have stated, the construction industry must start copying best practice from other business sectors and thus ought to start regularly measuring the things that all other business sectors have long believed are leading indicators of improvement. The following definition sums up the view of what improvement is about in all other sectors and points clearly at what aspects of performance should be regularly measured to test the rate of improvement in the construction industry, namely that:

'Real improvement in performance comes from the elimination of unnecessary costs (in the form of the inefficient utilisation of labour and materials) from the underlying labour and materials costs.'

The key lesson from all other business sectors is that unless the root cause of poor performance is regularly measured and targeted in the continuous improvement programme that runs through all the firms in the supply chain, any improvement will be superficial and transitory. As can be seen from the above definition, in all other business sectors the root cause of poor performance is the inefficient utilisation of labour and materials. Consequently, the construction industry ought to logically replicate what is done in other business sectors, accept that the root cause of its poor performance is its inefficient utilisation of labour and materials and therefore start to measure its inefficient utilisation of labour and materials.

After all:

❑ If the strategic partners do not know how much disruption is regularly encountered on site, how can they know their performance is improving?
❑ If they do not know how much rework regularly occurs (including rework by the professional designers), how can they know their performance is improving?
❑ If they do not know how often materials or plant arrive late, how can they know their performance is improving?
❑ If they do not know how often cluttered work areas cause delay and disruption, how can they know their performance is improving?
❑ If they do not know how much disruption and reworking is caused by errors on the drawings, how can they know their performance is improving?
❑ If they do not know the amount of new materials and components that are wasted, how can they know their performance is improving?
❑ If they do not know the amount of defective materials and components that have to be sent back for replacement, how can they know their performance is improving?

If the strategic partners are to implement fully the new collaborative working practices that are the basis of supply chain management and lean thinking, their leaders must inform themselves about their effective utilisation of labour and materials, since these make up over 80% of the total cost of construction. After all, what is lean thinking but the elimination of unnecessary costs?

Above all, the strategic supply-side partners need to learn from other business sectors that the responsibility for driving forward radical improvement is theirs, not the demand-side clients'. All the demand-side clients should do is to buy their buildings and constructed products the way they buy all other goods. The demand-side customer should merely buy the product that offers the best value for money, and this means buying a product where the supply-side supply chain is constantly improving its processes to reduce unnecessary costs to the absolute minimum, whilst driving up quality to a level that attracts the most customers.

7 The Client's Role in Partnering

As I said in the last chapter, in order to understand what the demand-side client's ideal role should be in partnering, it makes sense to look at the nature and size of the many and various demand-side customers in the construction industry and then compare them with the nature and size of demand-side customers in other business sectors, and to look at the role of customers in those other sectors. This is also the approach advocated by virtually all of the various reports on improving the performance of the construction industry from across the developed world. They insist that the construction industry stops seeing itself as different to all other business sectors and starts importing the tools and techniques that those other business sectors have used to drive forward improvements in performance over the last few decades.

It is also necessary for demand-side clients to understand the enabling role that is constantly advocated for them in various reports on improving the efficiency of the construction industry. This role is about buying their buildings and constructed products in the same way that they buy all their other goods. In short, they ought to procure their building in the same way that they would procure a new car, a new computer or a new shirt.

This puts partnering where it should be, on the supply-side of the interface, because in those other business sectors partnering is primarily about the close, collaborative, long-term, strategic relationships between the supply-side members of the design and manufacturing team. In the case of a new car, the demand-side customer would only consider some form of partnering if the procurement involved a fleet of cars.

There is a tendency in the UK to fall into the trap of assuming that all demand-side customers are large, repeat customers who are able to offer a continuous flow of medium to large projects. This is probably because these major customers are better able to make their voices heard and are better able to make senior staff available to join the key industry bodies. Small and occasional customers, however, have neither the time nor the inclination to get involved with the key industry bodies, nor the status to make themselves heard. This false assumption about the true nature of the demand-side customers has caused the industry to believe that partnering can deliver significantly better value, even when it relates solely to the demand-side customer's relationship with the industry and the industry itself remains as fragmented and adversarial as it always has been.

Unfortunately, in numerical terms the bulk of demand-side customers are small and occasional customers who only have a one-off or intermittent requirement for a building. In reality, there are only a comparatively small number of major demand-side clients in a position to partner with one or more supply-side teams on a relatively constant flow of medium to large projects.

A good example of a typical intermittent demand-side customer is the large number of UK district councils, where the size and nature of their property portfolio is such that there may well be a gap of several years between construction projects. I often speak to staff at the sharp end in district councils and they find it extremely difficult to understand why, or how, they should partner with the supply-side team on a single project. In fact, they correctly perceive the terms 'partnering' and 'single project' to be contradictory. They see little advantage to be gained from a partnering relationship with one or more supply-side teams when the gap of several years between successive projects means that there is no possibility of the individuals involved on the first project being available to come together on the next project. With such an intermittent flow of projects, there is no possibility of setting up a continuous improvement process whereby the individuals in the supply chain can continue to build on the lessons learned on successive projects.

In fact, it is fairly obvious that the range of demand-side customers in the construction sector is remarkably similar to the range of demand-side customers in the automotive sector. There are relatively few demand-side customers who are able to offer repeat purchasing partnering relationships with car manufacturers because they operate fleets of vehicles. The bulk of demand-side customers buy a single car every few years – just like the district councils do in the construction sector.

The construction industry's answer to this dilemma of most demand-side customers being small and occasional is to continue to insist that partnering is the answer to the industry's ills, but that partnering can deliver vastly better value even if it is restricted to the demand-side/supply-side interface on a single project. If this were true, surely other business sectors (where there is a long track record in effective supply chain management, lean thinking and partnering stretching back over several decades) would already be operating such single purchase partnering relationships? Yet I have no recollection of VW or Toyota suggesting to their one-off car buyers that they could get far better value if they entered into a partnering relationship with them.

Frankly, as far as construction is concerned, the assumption that improvement of the whole industry can be delivered through supply-side partnering with demand-side customers is simply untrue. It merely demonstrates misunderstanding of the purpose of supply chain management or of lean thinking, a pronounced reluctance to learn from other business sectors and a strong desire to maintain the fragmented and adversial *status quo*.

The role the demand-side customers ought to adopt was very well summed up in the *Accelerating Change* report in the UK in 2002, which said:

'Clients should require the use of integrated teams and long-term supply chains and actively participate in their creation.'

This demand in the *Accelerating Change* report does not require demand-side customers to partner with supply-side teams. Instead it requires them to:

Stop:
❑ Blocking supply-side integration by commissioning design team members separately from construction team members.
❑ Turning a blind eye to the way that the construction contractors appoint their sub-contractors and suppliers on a project-by-project basis using lowest price as the means of selection.
❑ Buying on the basis of lowest capital cost with no regard for the whole-life cost.
❑ Ignoring the magnitude of unnecessary costs in the underlying labour and materials costs.
❑ Ignoring the evidence from other business sectors that better value can only be delivered by the use of measurement to improve the utilisation of labour and materials.
❑ Blocking long-term, strategic supply-side partnering by interfering in the way the supply-side design and construction teams are formed, or by interfering in their selection of components and materials.

Start:
❑ Insisting that they will only do business with supply-side design and construction teams that can prove they are

already working together as a virtual company in which they are bound together within existing long-term, strategic supply-side partnerships.

❏ Demanding evidence of improvement from the measurement of the effective utilisation of labour and materials.

❏ Demanding evidence that the specialist suppliers (subcontractors, trades contractors and manufacturers) within the virtual company are always fully involved in design development from the outset.

❏ Using selection criteria that eliminate those supply-side design and construction teams that have not actively embraced supply chain management, lean thinking, a whole-life approach to costs and long-term, strategic supply-side partnering.

❏ Ensuring the use of proven experts in supply chain management and lean thinking to properly train all staff and ensure that everyone is walking the talk.

Figure 7.1 illustrates the demand-side client's relationship with the supply-side design and construction team. The main construction contractor (or lead firm in the supply-side virtual company) needs to have long-term, strategic partnering relationships with all the members of the supply-side design and construction team, since these are essential to the long-term continuous improvement process that reapplies lessons learned on successive projects and thus continuously reduces unnecessary costs and enhances quality. In fact, if these long-term, strategic supply-side relationships are not already in place when the demand-side client is considering tender long lists, it would be wise to discard the firm in question because it would be unlikely to be able to deliver value for money.

However, Figure 7.1 also illustrates that the client does not need to be an intrinsic part of these long-term, strategic supply-side partnering relationships that underpin the continuous improvement process. Figure 7.1 shows that the

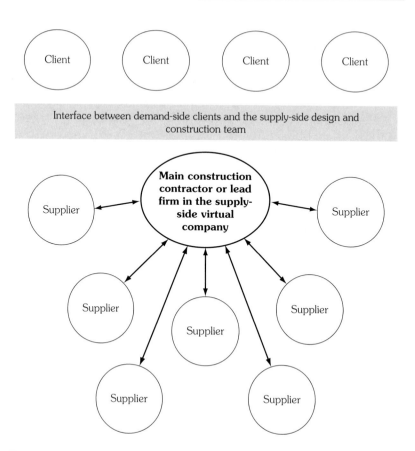

Figure 7.1 Partnering in the construction sector

client's role should be to interface closely with the supply-side design and construction team for the duration of the specific project to ensure that the team fully understands every aspect of the end user's business and functional requirements.

It may be that on occasions major demand-side clients have a succession of projects that follow closely one after another and would therefore lend themselves to a long-term partnering relationship with a specific design and construction team. However, such demand-side/supply-side partnering relationships ought to be predicated on the pre-existence of partnerships that are able to provide hard evidence from accurate measurement that they have consistently reduced unnecessary costs on a year-by-year basis, and that this reduction in unnecessary costs has been used to reduce both initial capital and whole-life costs. Thus, in every situation, the supply-side partnering relationships are of primary importance and should pre-date any relationship with the demand-side client. This reflects best practice in supply chain management and lean thinking in other business sectors, as can be seen very clearly if Figure 7.1 is compared with Figure 1.1 in Chapter 1.

Clearly, the demand-side customers have a key role to play in the reform of the construction industry, but this role is not about them partnering with supply-side firms that continue to operate in the same fragmented and adversarial ways that have caused the industry's inefficiencies in the past. The demand-side customer's role is to change their procurement practices to:

❑ force the total integration of the entire supply-side design and construction team;
❑ force the introduction of long-term, strategic supply-side partnering relationships;
❑ force the supply-side to measure its effective utilisation of labour and materials;
❑ force the supply-side to think in terms of whole-life cost rather than lowest price.

In fact, it is almost certain that unless the demand-side customers change their procurement practices to a form that forces, encourages or facilitates integration of the design and construction supply chain through long-term, strategic supply-side partnering and the delivery of the highest optimum whole-life performance for the lowest optimum whole-life cost, the current drive for improvement will quickly die away. If the majority of demand-side clients started to insist that they would only procure their buildings and constructed products from virtual companies where the design and construction supply chain firms (including trades contractors and manufacturers) were able to prove the existence of long-term, strategic supply-side partnerships that were based on improving performance through the elimination of unnecessary costs, so that they could deliver constructed facilities to their demand-side customers for the lowest whole-life cost and highest whole-life quality, I find it difficult to believe that the industry would not change its act overnight.

Earlier chapters discuss the six guidelines that the UK Rethinking Construction organisation listed in their publication *Rethinking the Construction Client – Guidelines for Construction Clients in the Public Sector*. The six guidelines define the targets that public sector clients seeking best value should aim for when procuring constructed products, but they are equally valid for private sector clients who are determined to achieve the best whole-life value for the lowest whole-life cost. The six guidelines are:

❑ *'Traditional processes of selection should be radically changed because they do not lead to the best value.*
❑ *An integrated team, which includes the client, should be formed before design and maintained throughout delivery.*
❑ *Contracts should lead to mutual benefit for all parties and be based on a target and whole-life cost approach.*
❑ *Suppliers should be selected by best value and not by lowest price: this can be achieved within EC and central government procurement guidelines.*
❑ *Performance measurement should be used to underpin continuous improvement within a collaborative working process.*
❑ *Culture and processes should be changed so that co-operative rather than confrontational working is achieved.'*

Earlier chapters also explain the consequences for the UK construction industry of the publication of the Confederation of Construction Clients *Charter Handbook*, the National Audit Office report *Modernising Construction* and the Department of Culture, Media and Sport report *Better Public Buildings*. The existence of these three publications, particularly *Modernising Construction* and *Better Public Buildings*, means that all public sector clients (who represent 40% of the total UK construction market) will inevitably have little option but to adopt the best-practice approach described in the National Audit Office and Department of Culture, Media and Sport reports. They are all subject to external audit by the National Audit Office (for central government procurers of construction services) or by the Audit Commission (health authority and local government procurers of construction services) and as these two audit bodies will inevitably use *Modernising Construction* and *Better Public Buildings* as the common evaluative criteria for all their assessments, those audited by them will have little option but to adopt the same approach to best practice.

The private sector will obviously have far greater freedom to choose an appropriate approach to procurement. However, if they choose to adopt the principles of the *Charter Handbook*, as the *Accelerating Change* report strongly advocates, their procurement practices will need to be reviewed against the six goals of construction best practice and changed where they diverge.

It is made clear in previous chapters that *Charter Handbook*, *Modernising Construction* and *Better Public Buildings* are very similar and throw up the same two key differentiators and the same six primary goals of construction best practice. These are as follows.

The two key differentiators of construction best practice

❑ '*Abandonment of lowest capital cost as the value comparator. This is replaced in the selection process with whole-life cost and functional performance as the value for money comparators. This means industry must predict, deliver and be measured on its ability to deliver maximum durability and functionality (which includes delighted end users).*

❑ **Involving specialist contractors and suppliers in design from the outset.** *This means abandoning all forms of traditional procurement which delay the appointment of the specialist suppliers (sub-contractors, specialist contractors and manufacturers) until the design is well advanced (most of the buildability problems on site are created in the first 20% of the design process). Traditional forms of sequential appointment are replaced with a requirement to appoint a totally integrated design and construction supply chain from the outset. This is only possible if the appointment of the integrated supply chain is through a single point of contact – precisely as it would be in the purchase of every other product from every other sector.'*

The six primary goals of construction best practice

❑ *'The finished building will ensure maximum functionality.*
❑ *The end users will benefit from the lowest cost of ownership.*
❑ *Inefficiency and waste in the utilisation of labour and materials will be eliminated.*
❑ *The specialist suppliers will be involved in design from the outset to achieve integration and buildability.*
❑ *The design and the construction of the building will be achieved through a single point of contact for the most effective co-ordination and clarity of responsibility.*
❑ *Current performance and improvement achievements will be established by measurement.'*

There is good case history evidence to prove that the adoption of the supply chain management principles that are common to the *Charter Handbook*, and to the *Modernising Construction* and the *Better Public Build-*

ings reports, will deliver major commercial improvements to both the demand and the supply side of the construction industry.

Key to these supply chain management techniques is long-term supply-side partnering within a virtual company in the form that has been the foundation of improved performance in other business sectors over the last few decades. This is very clearly explained and emphasised in Chapter 2, where the following quotation is given from the SCRIA handbook that had been developed by Sigma Management Development Ltd on behalf of the UK Department of Trade and Industry to facilitate radical improvement of the UK aerospace industry:

'*The enlightened companies have recognised that for the supply chain to work to its optimum, the flow of information has to be excellent. They have selected a sub-set of suppliers with whom they form closer relationships in order to facilitate the information flow. These closer relationships can be regarded as a form of partnership. The key to "oiling the wheels" of the supply chain is for companies to decide which suppliers and customers have the potential to add most value to their business and agree a form of partnership. There are no pre-defined rules for a partnership. Each one should be constructed so as to be appropriate to the relationship required.*'

Not only does the adoption of supply chain management principles and long-term, strategic supply-side partnering deliver improved lifetime quality and functionality and reduced capital and operating costs to the end user, but also the profits made by the supply-side firms will be enhanced by converting inefficiency and waste in the utilisation of labour and materials into far lower prices and far higher profits.

The St Helens and the Building Down Barriers case histories I describe later in this chapter provide hard evidence of the considerably better value that can be achieved if the traditional lowest price tendering is replaced by a selection process that focuses wholly or primarily on the experience, knowledge and skills of a fully integrated supply chain, especially those of the specialist suppliers (sub-contractors, trades contractors and manufacturers) and ensures (preferably enforces) the involvement of the specialist suppliers in design from the outset of the development of the design. The two case histories also show the magnitude of the improvements that can be delivered if the selection process also demands hard evidence to back up all the claims made about improved performance by the competing design and construction teams.

The St Helens Metropolitan Borough Council approach to best value selection came out of recognition that lowest price tendering regularly failed to deliver best value. Out-turn prices regularly exceeded tender prices by a significant amount, buildings were regularly delivered late, maintenance costs were never predicted and regularly came out higher than expected, components and materials regularly failed earlier than expected, and end users were regularly dissatisfied with the functional performance of the completed buildings.

Since lowest price tendering regularly failed to deliver best value, St Helens (officers, elected members and internal audit staff) took the very courageous decision to abandon lowest price tendering. They decided that best value was most likely to be delivered if their selection process selected the supply-side team that could demonstrate, and prove with hard evidence, that it had the highest level of appropriate skills, experience and knowledge.

The demand-side client would announce and set aside from the selection process the project's gross maximum price (which generally came from externally imposed government cost limits). The selection process could then focus wholly on the skill and experience of the supply-side team by demanding evidence of the use of an open-book approach down through the supply chain, of effective supply chain management, of total integration of the design and construction supply chain, of effective application of value management and value engineering techniques, and of the involvement of end users in the value management and value engineering workshops. The selection process would also predetermine the mechanism by which savings would be shared between the end-user client and the supply-side team on a 50–50 basis.

The St Helens case history proves how effective their best value selection process was in the first project. There have now been many other projects for a variety of local authorities and public sector clients based on the best value selection process, including several housing association projects, and all have demonstrated beyond any possible doubt that the approach out-performs all lowest price tendering processes.

In fact, the primary school in the case history was compared with another primary school close by, where the contract had been let shortly before the case history school and had been awarded on the basis of the usual lowest price tendering approach. Both were identical in size and functional requirement. The out-turn price (final account) of the other primary school exceeded the tender price by around 11%. In addition, the tender price of the other school exceeded the gross maximum price of the case history school by around 9%, but the gap between the two schools was further widened because the best value approach delivered the case history school at around 18% less than the gross maximum price.

This meant that the case history school that used the best value selection process was around 38% better value than the school that used lowest price tendering. In fact, the case history school was delivered with far better functionality than the other school because the supply-side team, and the Headmaster of the school (with St Helens Metropolitan Borough Council's agreement) agreed that the 18% would be ploughed back into the school to give even better value.

The original St Helens Metropolitan Borough Council best value selection process has subsequently been refined and improved in its usage by an ever-increasing number of local authorities and housing associations across the north of England, which have been coached and trained in its usage by the ex-Deputy Chief Executive from St Helens, who first developed the process and is now operating as a consultant

---

Case History No. 1 – St Helens Metropolitan Borough Council

The Borough Council introduced a 'best value' approach to construction tendering as an alternative to the traditional confrontational approach of the industry with its 'sealed envelope' lowest price process, which failed to provide clients with value for money because projects would regularly exceed the original cost and completion time predictions.

Their best value tendering approach appointed the entire design and construction supply chain from the outset, had a selection process that was based solely on the skills and experience of the entire design and construction team (including the specialist suppliers) rather than the traditional 'lowest price' approach, and included sharing savings with the supply-side team on a 50–50 basis.

When tested on a £2.3 million primary school, the out-turn cost to the council remained the same as the original cost estimate but enhanced standards amounting to £400 000 were able to be built into the project.

The school was completed ahead of the original planned completion date, much higher mechanical and electrical standards than the norm were achieved and the site had an excellent safety record. The headmaster and his staff were involved in the partnering process from day one and were convinced that functionality was well above that usually delivered by new buildings.

> **Case History No. 2 – Building Down Barriers pilot projects**
>
> The two pilot projects used as test-beds for the Building Down Barriers supply chain management system produced exceptional improvements in performance over conventionally procured buildings.
>
> Labour efficiency on site reached 70%, which was almost double that usually achieved, and both specialist supplier teams came close to achieving a 'right first time' culture. Wastage of new materials was close to zero, whereas the industry norm is around 30%.
>
> Both buildings were completed ahead of the contract completion date, with one being completed 20% ahead. The through-life cost forecast for both buildings was well below that for traditionally procured buildings, without incurring a significantly increased capital cost.
>
> Functionality was demonstrated at every stage of design using three-dimensional visualisation, which avoided the usual problem of end users struggling to understand two-dimensional drawings, and the end users were delighted with actual functionality of the completed buildings, which exceeded anything they had experienced before.

on best value tendering. The best value tendering process has developed into a two-part process that uses carefully structured questionnaires marked by an in-house team to avoid any possible risk of accusations of subjectivity.

The selection process has also been developed to favour those supply-side teams that are already using long-term, strategic supply-side partnering to support the delivery of far better value to the demand-side customers. This is because considerable experience in the usage of the best value process has very clearly demonstrated that best value (in terms of lowest capital cost, best whole-life performance and best functional performance) is most likely to be delivered where the design and construction supply-side teams have been working together on previous projects (for a variety of other clients) within long-term, strategic supply-side partnerships or alliances that are targeted at the reduction of unnecessary costs and the delivery of best value to the end-user client.

This point is emphasised at the beginning of this chapter, where I strongly advise that demand-side/supply-side partnering relationships ought to be based on the pre-existence of long-term, strategic supply-side partnerships able to provide hard evidence from accurate measurement that they have consistently reduced unnecessary costs on a year-by-year basis, and that this reduction in unnecessary costs has been used to reduce both initial capital costs and whole-life costs. The considerable experience of supply chain management and lean thinking in other business sectors has taught us that long-term, strategic supply-side partnering relationships are of primary importance and should pre-date any relationship with the demand-side client.

The stage one questionnaire now in use by those local authorities and housing associations that are adopting the current version of the best value tendering process asks a range of questions that includes the following:

- ❑ Can you demonstrate partnering experience with other members of your supply chain?
- ❑ Do you benchmark against Construction Best Practice Programme Key Performance Indicators, or other internal Key Performance Indicators?
- ❑ Do you benchmark your performance through post-contract (post-handover) questionnaires, including client/customer satisfaction feedback?
- ❑ Do you encourage training through the Construction Skills Certification Scheme, or other recognised schemes?
- ❑ Can you provide any evidence of your firm's commitment to change under Rethinking Construction?
- ❑ Can you demonstrate areas of innovation and good practice introduced by your firm?

❑ Can you demonstrate how your firm is dealing with the training necessary within your organisation to deal with the culture change required?

❑ List five key skills that you would expect your own and your supply chain's key personnel to possess as a group.

❑ What increased productivity do you envisage will ensue from good partnering practices?

❑ Please list up to six mechanisms that you use to ensure effective management of time scales within projects.

❑ Give examples of how you identify best practice and how the information is used in your process of continual improvement.

❑ Explain what 'Respect for People' means in your organisation, outline its application and benefits to you and your supply chain partners.

The marking system reduces the responses to the stage one questionnaire down to five or six firms that are then asked to complete a project-specific questionnaire that seeks method statements setting out how the supply-side team would approach the specific project.

This includes the following:

❑ **The competency, understanding and skills of the supply-side team.** These are required to relate to the areas of expertise highlighted by the stage one questionnaire.

❑ **Ability to change.** This is related to the mechanism by which the supply-side team would apply well established improvement processes (which includes changing the culture of the supply-side firms) to improving the culture of any local firms that were introduced into the supply-side team to meet the council's requirement to make maximum use of local labour.

❑ **Value management/value engineering.** This requires the method statement for the application of value management techniques to define the precise end-user functional requirements and the method statement for the application of value engineering techniques to drive out unnecessary costs in the use of labour and materials and to drive up whole-life quality.

❑ **Systems management.** This requires the method statement for the management of all the design and construction processes and activities so that 'right first time' would be achieved on site.

❑ **Best practice and innovation.** This requires the method statement that shows how best practice and innovation would be imported into the design and construction process for the specific project. This is especially important where new firms are introduced into the supply-side team to meet the council's requirement to make maximum use of local labour.

❑ **Pre-/post-contract costs.** This requires the method statement that sets out the management system that would control the underlying labour and material costs throughout the design and construction supply chain, especially the capturing and sharing of savings and the management of risks.

The two case histories and the best value tendering process emphasise the importance to the demand-side client of delivering the precise functional performance required by the end user, of using value management techniques to accurately capture and record the end user's functional requirements and of involving the end users in value engineering workshops. The critical importance of delivering a level of functional performance that delighted the end users was also a key theme of the *Charter Handbook*.

As a consequence, the demand-side client's selection process must (as the latest version of the best value selection process does) demand evidence from the supply-side team of the method they have used on previous projects to deliver maximum functionality. The method statement must include the value management techniques that were used to capture and record the end user's specific functional requirements, the value engineering techniques that ensured that the functional requirements were fully incorporated into the developing design, and the measurement system that was used to test the functional effectiveness of the completed building.

Interestingly, the critical importance to the end users of the efficient functional performance of the completed building is not unique to the UK. The same concern can be found in virtually all developed countries and is expressed very clearly in the 1996 report that sets the US National Construction Goals. The US report makes the point that if the salaries of the occupants are included, the running costs of a typical office block for a single year equal the capital cost of construction. In fact, the US *National Construction Goals* report includes a target for better design to deliver a 30% improvement in occupant efficiency. Since it is unlikely that design in the UK is delivering significantly higher occupant efficiency levels than in the USA, a similar improvement in UK occupant efficiency ought to be possible if the integrated approach described in all three UK standards (*Charter Handbook, Modernising Construction* and *Better Public Buildings*, see Chapter 3) is adopted by the UK construction industry.

The focus on delivering occupant efficiency is reflected strongly in the two key differentiators and the six primary goals of best practice listed at the beginning of this chapter. It does not require a great deal of logical thought to recognise the commercial benefit to the end users of buildings that enable the occupants to boost their efficiency by 30%, especially where the functional excellence and the environment is such that the morale of the end users is raised to a level where they are delighted to be working in, or using the building. The UK *Better Public Buildings* report makes the very telling point that if hospitals provided a functionally excellent environment, patients would recover more quickly and would therefore be discharged earlier and this would represent a major operational saving for the hospital.

Both case histories also demonstrate that the use of performance measurement techniques within a continuous improvement regime (that can only happen through long-term, strategic supply-side partnering) could dramatically increase the effective utilisation of labour and materials. The evidence from the two case histories fully validates the claim made in the 1994 Latham Report that inefficiency and waste in the utilisation of labour and materials consumes 30% of the initial capital cost of construction.

Since it is highly unlikely that inefficiency in the UK is restricted to new construction activities, it is almost certain that construction activities on maintenance and refurbishment in the UK are also suffering a level of unnecessary cost due to the inefficient utilisation of labour and materials that amounts to 30% of the initial capital cost.

It follows from the above that there is an urgent and overwhelming need for a radical change in the sequential, fragmented procurement practice of clients and end users. All three standards (*Charter Handbook, Modernising Construction* and *Better Public Buildings*, see Chapter 3) see the consequences of sequential procurement as the primary cause of poor value from the construction industry.

Demand-side clients need to follow the courageous example of such clients as St Helens Metropolitan Borough Council and take the lead by replacing their lowest price tendering systems with best value selection that focuses on skills rather than price. The pace of reform of the construction industry in the UK would be dramatically increased if far more demand-side clients followed the St Helens lead and used a similar best value selection system that strongly favours those supply-side teams that have already embraced long-term, strategic supply-side partnering and formed themselves into virtual companies.

In the case of public sector clients, the widespread take-up by local authorities and housing associations of a best value selection system derived from the St Helens Metropolitan Borough Council system has been proven beyond doubt and can be copied by any public sector client. In fact, since St Helens has now been publicised by the UK government as a Beacon local authority, it now has a duty to help other local authorities copy its system.

For those clients who wish to develop their own value-based selection system, there is still a need to adopt a degree of commonality with other clients to avoid the consequences of a confused free-for-all, where each client adopts a different selection system and the construction industry has to bear the cost of having to interpret, and respond differently to, each client's differing selection system.

As Rethinking Construction has given a lead to clients by adopting and promulgating the six goals of construction best practice that were first listed in the Construction Best Practice Programme booklet *A Guide to Best Practice in Construction Procurement*, it would make sense for demand-side clients also to adopt the same six goals and make them the basis of their selection system. Thus, demand-side clients who wish to abandon lowest price tendering and are determined to introduce a value-based selection system could start the change process by willingly and enthusiastically embracing the two differentiators and the six goals of best practice procurement that are the foundation of the three standards.

This also applies to those professional advisors who provide an interface between the end users and the construction industry, whether employed internally by the client within a property division or employed externally as consultants, since they normally provide the expert advice on procurement methods and, quite often, they place the design and construction contracts with the industry. Frequently, it is the client or end user's professional advisors who are responsible for the sequential appointment of the consultant designers and the construction contractors. Consequently, their actions and advice to the client or end user make it impossible for the specialist suppliers to have any input into the initial concept design, even though it is well known that most of the buildability problems on site are created in the first 20% of the design process.

One aspect of traditional procurement practice of some clients that directly inhibits supply chain integration and the formation of long-term, strategic supply-side partnering is the practice of treating design as an in-house function that can safely be separated from construction. Where this occurs, it is almost impossible for the specialist suppliers to be involved in design from the outset and thus it is almost impossible to harness their experience and skill to eliminate inefficiency and waste in the utilisation of labour and materials and thus achieve a 'right first time' culture on site. Thus, where in-house design continues to be retained by the client, it is very unlikely that the client will ever achieve best value in construction procurement and the demand-side client's procurement practices will act as a major barrier to effective supply chain management and successful partnering.

The manner in which the best practice of the three standards is introduced by clients will be different, depending on the nature of each client's construction programme. The majority of clients are small and occasional procurers of construction and therefore are not in a position to become a major driver of the change process. The clients who are repeat procurers of construction have considerable leverage and can use their changed procurement practice to force the pace and direction of the change process. Excellent examples of this in the UK are BAA with their framework contract approach, Defence Estates with prime contracting, and NHS Estates with their Principle Supply Chain Manager approach. All three involve the introduction of supply chain integration and management by means that are appropriate to each client and all three believe that involving specialist suppliers in design from the outset, and thus harnessing their considerable skill and experience while the design solution is still at concept stage, will deliver better value.

The following sections briefly describe a possible internal change process that could be adopted by repeat clients and by small and occasional clients that are determined to change their existing procurement practices so that they abandon lowest price tendering and embrace value-based selection. It should be read bearing in mind the St Helens Metropolitan Borough Council and the Building Down Barriers pilot project case histories described earlier in this chapter and the best value selection system that was described in some detail immediately following the case histories. The suggested internal change process is as follows.

INTERNAL CHANGE PROCESS FOR DEMAND-SIDE CLIENTS TO REPLACE LOWEST PRICE TENDERING WITH VALUE-BASED SELECTION

Repeat clients

Assess whether you want the cost benefits of the construction best practice of the six primary goals, such as considerably lower capital and operating costs. These come from harnessing the very considerable knowledge and experience of the specialist suppliers from the outset of design development in order to eliminate inefficiency and waste in labour and materials utilisation, and from the use of high quality and durable materials and components. Construction best practice should provide accurately predicted, significantly lower and risk-free running costs that can confidently be built into long-term business plans. Construction best practice also delivers reduced operating costs that come from the impact that excellent functionality has on the efficiency and morale of the workforce. Construction best practice delivers greater cost and time certainty that comes from the actual out-turn cost never exceeding the forecast and the building always being completed on time.

   To assess how well your current design and construction process accords with best practice, check if your construction contractors measure the efficient utilisation of labour and materials throughout their site activities. Speak to specialist suppliers (sub-contractors and trades contractors) about the frequency and causation of disruption to their work on site. Undertake end user surveys to test the effective functional performance of completed buildings. Compare out-turn costs and actual completion dates with the initial budget and the forecast completion dates of a selection of completed buildings. Seek advice from your facility management staff on the maintainability, durability and ease of operation of completed buildings and facilities.

If the results of this assessment reveal a gap between your current design and construction process and the best practice of the six goals, you would be wise to adopt the following action plan to change your internal procurement working practices:

❑ Compare current practices and processes with the two key differentiators and the six primary goals of construction best practice, i.e. does the current procurement process ensure the involvement of specialist suppliers from the outset of the design process, which is critical to a 'right first time' culture on site? If current practice starts with the appointment of an architect to develop the design, who then appoints a construction contractor, who in turn appoints specialist suppliers at a time when it is far too late for them to have any influence on the design (especially the elimination of inefficiency and waste in the utilisation of labour and materials), it is unlikely that the six primary goals will be delivered and you are inevitably paying a heavy cost penalty. Always ensure that this comparison is done objectively and never make assumptions or accept anyone's opinion as fact. It may be best to seek the assistance of an independent expert, who can demonstrate a comprehensive understanding of the six goals, to assist the comparative analysis.

❑ Once the comparative analysis has objectively established any variance between current practices and best

practice, the next step is to set an improvement target
for the organisation and to ensure everyone fully understands what the target means, why it is important to the
organisation, what benefits the improvements will deliver, how the improvements will be measured, and
how the target will affect each employee's role.

❑ Set Key Performance Indicators that enable you to accurately measure the organisation's rate of improvement, such as: end-user satisfaction with functionality;
the production of accurate cost of ownership predictions
by the design and construction teams; the rate of improvement in the on-site utilisation of labour and materials; the stage when specialist suppliers actually become
involved in design development (especially whether their
skill, knowledge and experience is really being used to
the greatest advantage by the designers); and the speed
of introduction of single point procurement.

❑ Ensure all leaders, especially the head of the organisation, become knowledgeable crusaders and champions
for best practice. This requires them to have a deep and
consistent understanding of the six primary goals and to
ensure that every word they utter, every action they take,
and everything they write reinforces the change process.

❑ Communicate the improvement targets and the
intended changes in current practices and processes to
everyone within your own organisation and in those
organisations with which you do business in the construction industry. This ensures that everyone (including
everyone with whom you interface) fully understands
where you are going, why you are going there and
how you intend to get there. This is particularly important for those with whom you interface, because they
need to work out how it affects their own practices and
processes. It is imperative that the language used is such
that everyone can understand the message, there must
be no ambiguities and there must be some way of
checking the understanding of the recipients. The onus

is always on the sender of the message to use the most appropriate language and the KISS (Keep It Simple, Stupid) principle should always apply.

❑ Only invite expressions of interest from those design and construction firms that are able to provide well documented evidence that they have practices and processes in place that ensure delivery of the six primary goals. In particular, demand the names of those firms (including the design and other consultants, but particularly including specialist suppliers) that can prove that they are already bound together within an integrated team in long-term, strategic supply-side supply chain partnerships.

❑ Focus your in-house resources on working with the end users to define the business needs and detailed functional requirements in terms that are suitable for measuring and evaluating the output performance of the built solution, leave designing the solution to the integrated design and construction team.

Occasional clients (large and small)

Assess whether you want the cost benefits of the construction best practice of the six goals, such as considerably lower capital and operating costs. These come from harnessing the considerable knowledge and experience of the specialist suppliers from the outset of design development in order to eliminate inefficiency and waste in labour and materials utilisation, and from the use of high quality and durable materials and components. Construction best practice should provide accurately predicted, significantly lower and risk-free running costs that can confidently be built into long-term business plans. Construction best practice also delivers reduced operating costs that come from the impact that excellent functionality has on the efficiency and morale of the workforce. Construction best practice delivers the greater cost and time certainty that comes from the actual out-turn cost never exceeding the forecast and the building always being completed on time.

Always demand evidence from design and construction firms to prove what has been achieved on other projects. Have the designers or the construction contractors measured the efficient utilisation of labour and materials throughout the site activities and can they provide you with the figures to back up their claims? Have they undertaken end-user surveys to test effective functionality and can they provide you with copies of representative surveys? Have they compared out-turn costs with the initial budgets and can they tell you how often the two match? Have they compared the actual completion dates with the original target completion dates and can they tell you how often the two match? Have they any evidence from facility management staff of the maintainability, durability and ease of operation of completed buildings and facilities?

If the results of this probing demonstrate that the design and construction firms are not able to prove they are consistently delivering the construction best practice of the six goals you would be wise to adopt the following action plan:

❑ Only invite expressions of interest from those firms that are able to provide well documented evidence that they have practices and processes in place that ensure delivery of the six primary goals. In particular, demand the names of those firms (including the design and other consultants, but particularly including specialist suppliers) that can prove they are already bound together within an integrated design and construction team in long-term, strategic supply chain partnerships.
❑ If professional advice is deemed necessary, ensure those offering the advice can prove their deep and comprehensive understanding of the six goals. They should also have a full understanding of the National Audit Office report *Modernising Construction*. (The Be organisa-

tion at Reading University and the Construction Best Practice Programme at the Building Research Establishment can both assist in advising the names of suitable consultants.)

❑ Ensure that you restrict your own organisation to defining the business needs and detailed functional requirements in terms that are suitable for measuring and evaluating the output performance of the business activities housed within the built solution. Do not be tempted to define your requirement in terms of built solutions because to do so will transfer risk to yourself if it does not perform as efficiently as it should.

Self-assessment questionnaire for use in the internal change process by demand-side clients who want to replace lowest price tendering with value-based selection

An effective change process must start by establishing accurately how well your current procurement process compares with the best practice to which you aspire. The following questions have been devised from my Building Down Barriers experience, with refinements from the experiences of the various demand-side clients that have adopted the St Helens Metropolitan Borough Council best value selection system, and are offered to assist the internal self-assessment process.

When answering each question against an individual goal, use the available evidence to assess the degree by which the criteria are met, i.e. 0%, 10%, 20%, 30%, etc. of the time. Then calculate the average percentage for the specific goal to show the gap in performance that your improvement process must close. Analysing the answers will show where the strengths and weaknesses of your current procurement process lie and will enable your change process to be targeted at the weakest areas of performance.

Functionality

- ❑ How frequently do the people involved in your procurement process demand that the design and construction teams use value management workshops (that include end users) to define and prioritise the detailed functional or business needs?
- ❑ How often do the people involved in your procurement process demand that structured and facilitated value engineering workshops are used by the design and construction teams to support and validate design decisions and the selection of all the components and materials?
- ❑ How many of the staff involved in your procurement process understand why maximum functionality would benefit the effectiveness and morale of end users?

Cost of ownership

- ❑ How many of the staff involved in your procurement process have read the Confederation of Construction Clients' publication *Whole Life Costing – A Client's Guide?*
- ❑ How many of the staff involved in your procurement process ensure that those involved in the design and construction process have read the *Component Life Manual* produced by the Building Performance Group (which is linked to the Housing Association Property Mutual insurance company, see Further Reading)?
- ❑ How frequently do the employees involved in your procurement process demand that the design and construction team accurately predict the cost of ownership?
- ❑ How well do the members of staff involved in your procurement process understand the difference between accurately predicting and merely estimating the cost of ownership?
- ❑ To what degree do staff involved in your procurement process concern themselves about defects during the usage of a building and do they understand how these

could adversely affect the end user's efficiency and well-being?

Inefficiency and waste

❑ How many of the staff involved in your procurement process have any experience of working with construction teams that have measured the effective utilisation of labour and materials?

❑ How many of the staff involved in your procurement process have read the National Audit Office report *Modernising Construction*, particularly what it has to say about the degree to which the industry measures the effective utilisation of labour and materials?

❑ How many of the staff involved in your procurement process understand why measuring the effective utilisation of labour and materials is key to the elimination of unnecessary costs?

Specialist suppliers

❑ How many of the staff involved in your procurement process have direct experience of the significant savings that come directly from the buildability that is injected by the early involvement of specialist suppliers in design?

❑ How many of the staff involved in your procurement process have read the Reading Construction Forum publication *Unlocking Specialist Potential* or the Building Services Research and Information Association Technical Note TN14/97 *Improving M & E Site Productivity*?

❑ How many of the staff involved in your procurement process are committed to demanding the early involvement of specialist suppliers in the design process in order to deliver 'right first time' on site?

Single point of contact

❑ How many of the staff involved in your procurement process have any direct experience of single point of contact procurement?

❏ How many of the staff involved in your procurement process believe that single point of contact is the only form of procurement that would ensure the involvement of specialist suppliers from the outset of design development and thus the delivery of 'right first time'?

Measurement

❏ How many of the staff involved in your procurement process understand why performance measurement is fundamental to the elimination of unnecessary costs?

❏ How many of the staff involved in your procurement process understand what is meant by the term 'unnecessary costs'?

❏ How many of the staff involved in your procurement process would agree with the DTI dictum that *'If you don't know how well you are doing, how do you know you are doing well?'*, and are therefore demanding that construction contractors measure their effective utilisation of labour and materials?

As I've said constantly in this and earlier books, demand-side clients who are determined to get better value from procurement, especially better value in whole-life cost terms, need to stop selecting the design team separately from the construction team. They need to replace lowest price tendering with some form of value-based selection that appoints the entirety of the design and construction supply chain from the outset on the basis of their proven skill and experience. This moves the selection process from a simplistic focus on lowest price to a much more sophisticated analysis of the skill and experience of each member of the design and construction supply chain. It also requires the introduction of a more open and trusting relationship between the client and the integrated supply-side team, founded on performance measurement and partnering philosophy.

In the UK, government policy is that procurement should be on the basis of value for money and not lowest cost. The UK Minister of State for Trade and Industry has stated:

'The obsession with getting the lowest price for construction projects wastes money and cheats communities. Lowest cost does not mean better value.'

The more enlightened repeat clients in the UK have recognised that lowest price tendering has all too often turned out to be high risk to the commercial effectiveness of their business, because of the unpredictable price escalation that invariably occurs between the tender stage and the final settlement stage. They are exploring alternative ways of selecting an integrated design and construction supply-side team, and of awarding the contract, that place more reliance on trust, team working, openness and the sharing of savings and risks. BAA, Argent and McDonald's are good examples of this in the UK private sector and Defence Estates and St Helens Metropolitan Borough Council are good examples in the UK public sector.

The private sector clients who are moving away from tendering based on the lowest price tend to be major retailers who are using their skill and experience of managing their retail supply chain to actively manage their design and construction supply chain through various forms of supply chain partnerships. Whilst this approach by major retailers is both valid and low risk for clients who possess a high level of supply chain management skill and experience from their retail or manufacturing side, it may be high risk and difficult for clients who lack this skill or cannot afford the skilled resources that would be necessary to actively manage the design and construction supply chain.

A reality of construction procurement is that in almost every case, the client and the end user have a budget for the construction works that represents the maximum affordable price for constructing a building or facility to house the relevant functional or business activities. This may come from the business planning process that is an essential part of assessing the affordability of the venture in a competitive market, or it may simply come from the amount of money available from savings, grants or a loan. In either case, if the final capital cost exceeds the maximum affordable price, the client will suffer some form of financial hardship and in the worst case (especially if the difference is over 50%, as the National Audit Office found it was in three out of four public sector projects) this may well cause serious damage to the client's business competitiveness. This is equally true of the long-term operational and maintenance costs: end-user clients cannot afford unpleasant and costly financial surprises at any stage in the operational life of the building or facility because they will inevitably have to be paid for out of their profit margin and will therefore adversely affect their commercial competitiveness.

To avoid the consequences of lowest price tendering, which can involve serious budget overruns and poor whole-life quality in terms of durability and inefficient functionality, the client needs to select an integrated design and construction team that is able to prove it has the skill and experience to deliver a construction solution that efficiently satisfies the client's business or functional needs and does not exceed the client's maximum affordable price.

The starting point for any repeat client who wishes to ensure their procurement of construction is consistently delivering best value in capital cost, whole-life cost and functional terms, is to measure how well their current procurement process has performed. To do this, the client needs to review the constructed outputs of the procurement process in terms of the final capital cost and the performance in use of the completed building or constructed facility. This requires an objective assessment of past construction works contracts to compare final settlements with tender prices. For example, how often did the final settlement match the tender price? By how much did the final settlement exceed the tender price in the worst case? What is the average escalation in cost between the tender price and the final settlement? How often has the construction team been required to measure the effective utilisation of labour and materials? What is the average level of the effective utilisation of labour and materials? It also requires an objective assessment of unexpectedly early component or materials failures during the planned life of the building or facility and an assessment of the impact on the commercial well-being of the end user of making good the failure. Finally, it requires an objective assessment of the functional performance of the building or constructed facility by carefully interrogating the end users.

The situation is obviously different for small and occasional clients, since they do not have a continuous stream of construction contracts with which to assess the effectiveness of their procurement process. Nevertheless, they would be well advised to ask their professional advisors about the true performance of the procurement approach they recommend. Similarly, they would be well advised to ask any potential construction contractor (including management contractors and design and build contractors) about their effectiveness in delivering value for money.

Questions that would quickly paint a true picture are as follows:

❑ How often have final settlements matched tender prices?
❑ Where final settlements have exceeded tender prices, what was the worst case?
❑ What has been the average escalation in cost between tender price and final settlement?
❑ How often is the effective utilisation of labour and materials measured?
❑ What is the average level of the effective utilisation of labour and materials in recent construction works?
❑ What evidence have they of the true durability (performance in use) of the components and materials used in completed buildings or facilities?

❏ How often have they provided an accurate prediction of the annual cost of ownership of a building or facility?

❏ What evidence have they of end-user satisfaction with the functional performance of their completed buildings or facilities?

Any consideration of the appropriate procurement approach ought to bear in mind that the UK Latham and Egan Reports both made clear that design must be fully integrated with construction if clients are to achieve better value from construction procurement. The Egan Report in particular emphasised that the route to better performance by the industry, and thus to better value for clients, was for clients to buy their built products in the same way they bought their manufactured products. This clearly requires design to be an intrinsic part of the supply side and to be totally integrated with construction using appropriate supply chain management techniques. Buying a built product in the same way as a manufactured product and use by the supply side of supply chain management techniques also requires the supply side to deal with the client through a single point of contact.

Whilst every demand-side client is free to develop its own value-based selection system, I suggest a method later in this chapter that builds on the St Helens approach to value-based tendering and could be used by any demand-side client to select a fully integrated supply-side design and construction team that is already bound together within a virtual company on the basis of long-term, supply-side partnering relationships.

Value-based selection differs radically from lowest price tendering, in that it either sets aside price completely or it constrains price to less than 20% of the available marks. The focus of value-based selection is on the evidence provided by the members of the supply-side team of their skill, knowledge and experience. It is not about them merely telling you what they have done, it is about them showing you the evidence to prove it.

The best supply-side team will be able to provide evidence of how effective its improvement process has been in driving out unnecessary costs caused by the ineffective utilisation of labour and materials. It will be able to provide evidence of the effectiveness of its value management, value engineering and risk management techniques on previous projects to deliver maximum functionality and the lowest optimum whole-life cost. It will be able to name those of its suppliers (sub-contractors, trades contractors and manufacturers) already involved in supply-side partnerships and will be able to give you the results of regular performance measurement from the continuous improvement targets that have been tied into the partnerships.

The suggested way forward in the introduction of value-based selection of the supply-side team is for the demand-side client to focus on the skill and experience of the supply-side team through the use of a carefully structured questionnaire. Since this is contrary to normal industry experience, it may be sensible for the demand-side client to emphasise to the industry that price is not part (or is not a significant part) of the selection process.

Ideally, the demand-side client should declare from the outset the maximum affordable price for the building or facility and should also make clear that the selected supply-side team will be required to develop a target cost that is within the maximum affordable price as the design is developed. If possible, the maximum affordable price (or gross maximum price) should be expressed in both initial capital cost and long-term operational cost terms, as the *Charter Handbook* emphasises that best practice clients are far more interested in best whole-life value than lowest capital cost. Declaring the maximum affordable price (maximum gross price) from the outset and then setting aside the development of the target price until the supply-side design and construction team has been appointed will lead to a far more accurate target price because it will be based on the completed design.

It is extremely difficult, if not impossible, for the supply-side design and construction team to establish an accurate target price for a specific building or facility before the design is fully developed and the ground conditions have been fully explored and converted into the actual, project-specific foundations. Without the developed design, all that can be deduced is a ballpark price based on data from similar buildings or facilities that have been constructed in recent years.

Unfortunately, such prices are only approximate and have to be heavily qualified (or loaded with contingency sums) because of the large number of unknowns (such as the ground conditions, the fabric of the building, the configuration and size of the building, the town planning constraints). Leaving the development of the target price until after the design has been developed eliminates the risk of price escalation caused by the unknowns that regularly cause the price to rise considerably during the design and construction process and regularly cause the final account (final out-turn price) to be far higher than the original estimate, or even the tender price. However, the experience of demand-side clients, such as St Helens Metropolitan Borough Council, that have opted for best value or value-based selection of the supply-side team is that the supply-side team has always been able to establish a target price that was well within the maximum affordable price (gross maximum price).

I am also aware of three projects where the supply-side team was able to persuade the demand-side client to adjust the maximum affordable price upwards slightly because the use of through-life costs in the value engineering process had clearly demonstrated that the use of better quality and more durable components could pay back the higher capital cost within a few years. This decision was considerably helped by the integration of end-user clients into the supply-side team, because end-user clients could immediately appreciate the cost of ownership benefit that would come from the use of the more durable components and could then quickly adjust their long-term business plans to check the long term affordability of the capital cost/operating cost adjustment.

Any demand-side client who is worried about the supposed risks of best value or value-based selection would be well advised to make contact with those demand-side clients who have already embraced this new way of working and can therefore offer hard evidence of the results in terms of price and quality. Several local authorities and housing associations have already embraced best value selection and are convinced from their experiences that it offers far better value for money than the traditional lowest price tendering. Several major private sector clients have also embraced a similar value-based selection approach and would be able to provide very convincing evidence to prove that value-based selection is better value than lowest price tendering. In the case of private sector clients, the simplest way to locate them would be through Be (previously the Design Build Foundation).

There are obviously other approaches to selecting the supply-side team by value that are equally valid, but no matter what approach is used, the client should always ensure that the process is based on the selection of fully integrated design and construction teams that are tied together by pre-existing, long-term, strategic supply-side partnering relationships that embody the seven principles of supply chain management (especially the primary lean thinking principle). This approach accords fully with the UK Egan reports *Rethinking Construction* and *Accelerating Change*, it also fully accords with the Canadian, Australian and Singaporean reports, respectively, *Achieving Excellence in Construction, Building and Construction Industries Supply Chain Project* and *Construction 21*.

If price must be part of the value-based selection process, it should never count for more than 20% of the available marks and a two-envelope approach should be used so that the price can be dealt with separately from the skill and experience of the integrated team. It would also be wise to assess and award marks for the skills, knowledge and experience of the competing supply-side teams before price is considered, since there is always the risk that the prices in the second envelopes might colour the marking of the skills element of the bids.

Ideally, the envelopes containing the prices should not be opened until the contents of the skills envelopes have been assessed and marked. The delivery of best value will always come from the selection of the best design and construction team (in particular the selection of the best specialist suppliers or trades contractors), thus the marking system used should award the majority of the available marks to the skill and experience of the integrated team and should always do so on the basis of the well documented evidence of performance each team submits.

As I said above, the evidence submitted by the integrated design and construction supply-side team must include their recent achievements in delivering best value and at the very least ought to answer the following:

- ❏ How often have final settlements matched tender prices?
- ❏ Where final settlements have exceeded tender prices, what was the worst case?
- ❏ What has been the average escalation in cost between tender price and final settlement?
- ❏ How often is the effective utilisation of labour and materials measured?
- ❏ What is the average level of the effective utilisation of labour and materials in recent construction works?
- ❏ What evidence have they of the true durability (performance in use) of the components and materials used in completed buildings or facilities?
- ❏ How often have they provided an accurate prediction of the annual cost of ownership of a building or facility?
- ❏ What evidence have they of end-user satisfaction with the functional performance of their completed buildings or facilities?

The above questions, which relate to the delivery of best value for other clients, would provide a minimum assurance that the selected team has a good track record in the necessary skills needed to deliver best value on the current contract.

However, where the achievement of best value is a commercial priority for the client, it might be wise to ask the supply-side team to respond to a much more probing set of questions that could give greater assurance of their ability to deliver best value in terms of whole-life performance and the elimination of unnecessary costs. Basing the questionnaire on the six primary goals of construction best practice would ensure that all the necessary skills were adequately covered and would bring a degree of uniformity to the selection process. Such a questionnaire is suggested below.

VALUE-BASED SELECTION OF A FULLY INTEGRATED DESIGN AND CONSTRUCTION SUPPLY-SIDE TEAM (OR VIRTUAL COMPANY)

In order to bring a degree of commonality across the industry to that part of the selection process that evaluates skill and experience of the entire supply-side design and construction team, it would make sense to include a series of questions that relate to the six primary goals of construction best practice from the Construction Best Practice Programme booklet *A Guide to Best Practice in Construction Procurement* (these became the six primary themes of construction best practice from the Rethinking Construction publication *Rethinking the Construction Client – Guidelines for Construction Clients in the Public Sector*).

It would also be essential to make the selection process as fair and objective as possible by basing the marking system upon the tangible evidence of experience, practice or performance, rather than upon anecdotal claims. The following questionnaire may help the development of a value-based selection system, although it may need to be adjusted depending on the nature or size of the individual construction works.

When dealing with the response to a question against an individual goal, use the available evidence to assess the degree by which the criteria are met by all those involved in the design and construction process, i.e. 0%, 10%, 20%, 30%, etc. of the time. When the responses to the individual questions have been assessed, take the average of the percentage awarded to each question to give the overall percentage compliance with the individual goal. When each goal has been dealt with in this manner, the assessment will show where the strengths and weaknesses of the competing design and construction teams lie and will enable the selection decision to be fair and objective.

Where the results of the value-based selection have to stand up to internal (or external) audit, it would be sensible to consult closely with the auditors when devising the assessment questionnaire and to get their agreement to its final form. It should always be borne in mind that the unsuccessful bidders have a right to be de-briefed on the results of the assessment so that they can learn from the experience and better understand what they need to do better next time.

A possible approach might be to issue the questionnaire to each competing design and construction team and ask them to submit a report that provides the response to each question, with examples of the evidence that supports the response. For instance, in the case of the first question about value management workshops under 'Functionality', they might state that the frequency was 20% of all projects over the last five years and include one or two value management workshop reports from representative projects with attendance lists for the workshops that include end users, designers, construction contractors and specialist suppliers. In the case of the first question about cost of ownership predictions under 'Cost of Ownership', they might state that the frequency was 5% of all projects over the last five years and include one or two cost of ownership predictions from representative projects.

It would be essential to make clear to all the supply-side design and construction teams asked to make submissions that the term 'design and construction team' in the questionnaire refers to the entire supply chain and must therefore include the specialist suppliers (sub-contractors, trades contractors and manufacturers) that will actually be carrying out the construction activities for the building or facility.

Functionality

❑ How frequently are structured and facilitated value management workshops that include end users, design professionals, construction contractors and specialist suppliers (trades contractors) used to define and prioritise the detailed functional or business needs?

❑ How frequently are structured and facilitated value engineering workshops used by the design and construction teams to support and validate all design decisions and the selection of all the components and materials?

❑ How frequently are end users brought into the integrated teams to be involved in value engineering workshops?

❑ How many of those involved in the design and construction team have received appropriate training to assist their understanding of the principles of the *Charter Handbook* and of the full implications of supply chain management, especially the primary lean thinking principle of supply chain management?

❑ How frequently have end-user surveys been used after occupation to test the satisfaction of the end users with the functional performance of the completed building?

Cost of ownership

❑ How frequently has the end-user client been provided with an accurate cost of ownership prediction?

❑ Are such predictions always developed using relevant guidance, such as the UK Confederation of Construction Clients' publication *Whole Life Costing – A Client's Guide* and the *Component Life Manual* produced by the Building Performance Group?

❑ Can evidence be provided to prove the design and construction team has received relevant training in predicting the cost of ownership?

Inefficiency and waste

❑ How frequently have the firms making up the design and construction team measured the effective utilisation of labour and materials?

❑ Can the firms making up the design and construction team provide evidence of relevant training in the measurement and elimination of the unnecessary costs that are generated by the ineffective utilisation of labour and materials?

❑ How do the firms making up the design and construction team (especially the specialist suppliers or trades contractors) measure improvements in productivity (especially their effective utilisation of labour and materials)?

Specialist suppliers

❑ How often are specialist suppliers (sub-contractors, trades contractors and manufacturers) involved in design from the outset?

❑ Have all the specialist suppliers in the design and construction team been selected using value-based rather than lowest price selection?

❑ Are all the firms in the supply-side design and construction team tied together by long-term, strategic supply-side partnering relationships?

❑ Can the construction contractor member of the supply-side team demonstrate a commitment to strategic long-term partnering with suppliers?

Single point of contact

❏ How many of the supply-side design and construction team have worked together on previous projects as an integrated supply-side team?

❏ Can all the firms in the supply-side team demonstrate a commitment to long-term, strategic supply-side partnering?

Measurement

❏ Do all the firms in the supply-side team benchmark their performance against the UK Construction Best Practice Programme Key Performance Indicators or other internal Key Performance Indicators?

❏ How often do the specialist suppliers in the supply-side team (sub-contractors, trades contractors and manufacturers) regularly record the actual man-hours charged to the project (including abortive time caused by all forms of disruption and delay) and the actual materials used (including wastage) and compare these with the forecast figures?

❏ How often do the firms in the design and construction supply-side team conduct a detailed analysis of all forms of delay, disruption and materials wastage, and assess the causes and work with other team members to seek ways of eliminating those causes?

❏ Do the firms in the design and construction supply-side team set improvement targets that are aimed at using measurement to reduce (and eventually eliminate) the unnecessary costs caused by all forms of delay, disruption and wastage in the utilisation of labour and materials?

❏ Do the firms in the design and construction supply-side team open their books to other team members and share detailed performance data, such as profits, overheads and the effective utilisation of labour and materials?

Finally, remember that the demand-side client's approach to procurement is key to the reform and improvement of the construction industry. Unless the demand-side customers change their procurement practices to a form that encourages, enforces and facilitates total integration of the entire design and construction supply chain by demanding hard evidence of pre-existing, long-term, strategic supply-side partnering relationships, the current fragmentation and the resulting poor performance will continue. Without the demand-side clients using their procurement practices to force the pace and direction of change, the construction industry will not be able to deliver the highest optimum whole-life performance for the lowest optimum whole-life cost, nor will profit margins significantly increase, and the current drive for a radical improvement in performance will quickly die away.

Above all, do not forget the insistence in the UK *Accelerating Change* report that states:

'Clients should require the use of integrated teams and long-term supply chains and actively participate in their creation.'

8 Effective Leadership

As I have constantly emphasised in this and my earlier books, radical change in the way a company conducts its business will not happen without the changes in working practices being driven firmly and visibly from the top. This was illustrated by the following case history from the aerospace industry where a CEO described the active role he had to adopt in order to drive through a radical improvement in performance.

Case History – Aerospace industry

The importance of overt, powerful and committed leadership can be illustrated by an actual example of a major international firm from the aerospace industry that achieved a dramatic improvement in performance in a remarkably short space of time. Specifically, the cost of production was reduced by 30%, the production time was reduced by 50%, whilst the outstanding whole-life quality and performance of the engines was maintained.

The Chief Executive likened his role to that of a Crusader king. He said he had to be constantly seen by every one of his troops to be leading the way forward into battle. Every move he made, every phrase he spoke and every word he wrote had to reinforce and clarify the changes in working practices he wanted the firm to make. He had to ensure that everyone in the firm (and he said this must literally include every last person employed in his firm) must understand where the firm was going, why it must go there, what would happen if it failed to reach its destination, and (most important) what each individual had to do differently to be part of the change process. He said it was imperative that the tea lady and the cleaners felt they were included in the change process. They must understand why the changes in working practices are commercially essential and they must want to be an active part of the change process.

He also said that it was important to recognise and reward those who were making the greatest contribution to the change process. Quite often, the reward need be no more than public recognition by the Chief Executive for their efforts (i.e. a personal letter from the Chief Executive that is also put into the firm's newsletter).

He also said it was essential to provide everyone with regular progress reports which explained how the improvements in performance were being measured and what was being achieved in the various parts of the firm.

Equally important was the need to expose and deal with those that were blocking and opposing the changes in working practices. He said you could be absolutely certain that the grapevine would ensure that everyone would be aware of the names of those that were trying to covertly block the changes. It was equally certain that everyone was watching to see if the Chief Executive was on the ball, would pick up on what was really happening and would take action to eliminate the covert blocking.

The need for the CEO and the senior management team to take this pro-active stance is also reinforced by a consistent message from those at the sharp end within the construction industry. They constantly warn that from their perspective the pace of change is severely hindered, and in many cases halted, because the leadership of the organisation is failing to set out clear goals for the change process, failing to ensure everyone is clear about what they should be doing differently, failing to explain the commercial logic behind the new working practices, and failing to demonstrate by their own actions and words their total commitment to the new ways of working.

In fact, the complaint I hear time and time again is that those at the top are very good at talking the talk and using the right buzzwords like 'partnering', 'supply chain management', 'lean thinking' and 'supply chain integration', but that they are not very good when it comes to practising what they preach. The constant complaint is that senior managers rarely explain the meaning or the implications of the buzzwords, or explain what effect the buzzwords will have on existing working practices, or change their own actions in accordance with them.

When I was in Defence Estates, my two colonels used to frequently emphasise the need for clear leadership from the top if a new goal were to be attained. The example they used to illustrate their concerns was (not surprisingly) a battle to attain a specific goal in a war situation. They pointed out that the troops under your command need to know the precise purpose of the intended attack (for example, which particular hill within the range of hills they could see they were going to attack), they need to know the precise location of the enemy forces they have to contend with (i.e. where the shells and bullets would come from), they need to know the likely armaments of the enemy (will they have to guard against rockets, shells or just bullets?), they need to know the precise role each platoon or company was required to follow (i.e. to avoid the danger of 'friendly fire'), and they need to know the precise timing of the attack (i.e. to avoid half the attacking force being left behind). If all this is clearly explained in terms that every officer (commissioned and non-commissioned) could readily understand, they will have a good chance of succeeding in the coming battle.

What is unacceptable, the colonels insisted, is for a commander to use vague terms and imprecise instructions about who would do what, how the attack would be conducted, how the enemy would be configured or armed, or when the attack would be mounted. Such vagueness and imprecision would inevitably result in damaged morale, confusion, chaos, a lot of dead and injured soldiers, and a lost battle.

As the ultimate leader of the company, the Chief Executive must ensure everyone in the organisation, and everyone in the organisations with which it has business links, understands what working practices are to change, how they are to change, why they are to change, when the changes must be implemented, how the improvements in performance will be measured and what commercial benefits the changes will deliver.

Whilst the change process must be initiated and led by the Chief Executive in person, it is also true that a major change in working practices will not happen without a great deal of concerted effort on the part of everyone in the organisation and an acceptance by everyone that they will all have to change the way they work to some degree, with some having to change profoundly. Radical change can easily be thwarted by the inherent and powerful inertia of established custom and practices, and by covert resistance at all levels, especially at senior management level.

It must be obvious to all that the change process is owned, structured and directed by the Chief Executive in person, whose every action and every spoken and written utterance constantly reinforces the direction and the urgency of the change. The experience of organisations that have successfully embraced change teaches that the change process needs to be focused into four key areas if it is to be successful. Without all four being in place and operating concurrently, the change will become nothing more than wishful thinking and will soon be forgotten and replaced by the next bright idea.

The four essential and interlinked ingredients of successful change are as follows:

❑ A clearly explained and rational goal that all can understand, with which everyone can identify, and which can be related to specific improvements in performance that can be measured and compared with previous performance.

❑ Committed, determined and visible leadership by the Chief Executive that leaves no-one in any doubt about where the change must take the organisation, why it is commercially necessary to go there, and the time-scale for implementing the change.

❑ A detailed and comprehensive action plan for the development and implementation of the changes in working practices which explains in simple, easy to understand language what must be done differently by every member of the organisation. The provision of adequate and appropriate training must support this for everyone who is required to operate in a different manner.

❑ A simple and easy to understand explanation of the commercial benefits that will be delivered by the changes in working practices. This is best expressed in terms that relate to improved product quality, improved efficiency in working practices, reduced waste in the production process (both labour and materials) and, most important, reduced costs, lower prices and increased profits.

In any organisation or business sector, major changes in working practices are extremely difficult to initiate and achieve. Evidence shows that it is rare if as many as 30% of the workforce are in favour of the changes where they affect their own working practices. Another 30% will fight against the changes (usually in covert ways that will be concealed from the Chief Executive) because they are afraid that the changes will adversely affect their status, their ability to perform, or their pay. The remainder will sit on the fence until they are convinced the changes are inevitable and beneficial.

Those secretly against the changes in working practices will include many at middle and senior management level, possibly including some at Board level. They are generally older and more experienced in the familiar, comfortable and trusted ways of working. Quite often, they are also, unsurprisingly, worried that they are going to find it difficult to learn the new and unfamiliar ways of working at their time of life, and that this is likely to mean that their position in the hierarchy of the organisation will suffer as the younger and more junior staff more quickly learn the new ways of working and see their superiors struggling to cope.

It is obvious from the experiences of other business sectors that two important ingredients of a successful improvement process are communication and education. It is unreasonable to expect people to embrace radical change unless it has been explained in language they can understand and has been illustrated by examples drawn from their own day-to-day working practices. It is equally unreasonable to expect people to embrace radical change unless they have been given proper training in the new working practices. It is not acceptable to assume the explanation is adequate without verifying that it has been understood at all levels (with the responsibility for selecting the most appropriate language being that of the sender of the message). Nor is it acceptable to assume that existing training is either adequate or appropriate without verifying these assumptions by talking to the people at the sharp end who are the recipients of the training. It is also imperative to measure whether the training has actually changed working practices, preferably by the use of an objective, bottom-up feedback mechanism that has a good track record of success.

Those charged with communicating within the change process (including the Chief Executive) should bear in mind that it has always been a reality of buying and selling that you can generally buy using your own language, but you can rarely sell unless you use the language of the potential buyer. Consequently, those 'selling' the message about radical change within an organisation need to use the language of those that need to 'buy' the message and this may require the message being differently phrased for different recipients.

An excellent tool to measure and test the rate of improvement and to provide well structured support for the change process is available in the UK in the form of the European Foundation for Quality Management (EFQM) Excellence Model. This has an outstanding track record of success across the private and the public sector, in large as well as small organisations, in the UK and across Europe. The 'Award Simulation' self-assessment system provides an excellent mechanism for providing a regular, annual, consistent and very objective measurement of the rate of improvement from the people at the sharp end of the organisation. It tells the Chief Executive and the Board what is really happening at the coal-face, and it provides the workforce with an opportunity to ensure that the truth about what is going wrong with current working practices will get to the Chief Executive and the Board without being filtered or massaged by middle and senior managers.

The EFQM Excellence Model also forces organisations to adopt a very structured approach to improvement by the use of nine interdependent criteria, namely:

☐ **Leadership**. How leaders develop and facilitate the achievement of the mission and vision, develop values for long-term success and implement these via appropriate actions and behaviours, and are personally involved in ensuring that the organisation's management system is developed and implemented.
☐ **People**. How the organisation manages, develops and releases the knowledge and full potential of its people at an individual, team-based and organisation-wide level, and plans these activities in order to support its policy and strategy and the effective operation of its processes.

❑ **Policy and Strategy.** How the organisation implements its mission and vision via a clear stakeholder-focused strategy, supported by relevant policies, plans, objectives, targets and processes.

❑ **Partnerships and Resources.** How the organisation plans and manages its external partnerships and internal resources in order to support its policy and strategy and the effective operation of its processes.

❑ **Processes.** How the organisation designs, manages and improves its processes to support its policy and strategy and fully satisfy, and generate increasing value for, its customers and other stakeholders.

❑ **People Results.** What the organisation is achieving in relation to its people.

❑ **Customer Results.** What the organisation is achieving in relation to its external customers.

❑ **Society Results.** What the organisation is achieving in relation to local, national and international society as appropriate.

❑ **Key Performance Results.** What the organisation is achieving in relation to its planned performance.

Of the nine EFQM Excellence Model criteria, the most important are Leadership, Processes, Customer Results and Key Performance Results and this is reflected in the marking system that gives the following share of the total marks available:

❑ **Leadership.** 10%
❑ **Processes.** 14%
❑ **Customer Results.** 20%
❑ **Key Performance Results.** 15%

At the start of this chapter, four key essentials of any successful change process were listed. These were as follows:

❑ A clearly defined and unambiguous goal for the change process.

- ❏ Committed leadership of the change process from the Chief Executive.
- ❏ Well defined and clearly described processes for the improved working practices.
- ❏ A clear explanation of the commercial benefits the change process will deliver.

Not surprisingly, these four essential ingredients have a close correlation with the four most important criteria from the EFQM Excellence Model. Leadership is about setting a clear and unambiguous goal for the organisation and communicating that goal to everyone in language each person can understand. Process is about providing everyone with a route map that enables each of them to modify or change their out-moded working practices for best practice consistently across the entire organisation. Customer Results and Key Performance Results are about objectively measuring the commercial benefits that are delivered by the changes in working practices.

The importance of powerful and committed leadership cannot be overstated, the overwhelming evidence from those organisations in every sector of private and public activity that have successfully embraced radical change is absolutely clear: no matter how much superficial enthusiasm there is for radical change, it will be stillborn unless the Chief Executive is seen to be totally committed to it. This commitment must include the Chief Executive giving a simple and straightforward explanation of the goal for the change process that must be expressed in terms that everyone (without exception) can understand, and there must be some way of testing the understanding at an individual level across the organisation.

Feedback from organisations that have successfully achieved radical improvement warns that making assumptions about what people understood can be very misleading. It also carries a high risk that whilst everyone outwardly claims to have understood the message, the hidden reality is that everyone goes off in different directions because their interpretation of the message has been slightly different. It is imperative that their understanding is tested and validated, and the message must be re-phrased in more appropriate language if it is discovered that it has been interpreted differently across the organisation.

The change will also be stillborn if it is unwisely assumed that everyone will be able to modify his or her entrenched traditional working practices without considerable and appropriate help and education. In the UK, education needs to be carefully crafted and structured to address the areas where the working practices of the organisation do not match the six goals of construction best practice defined by the three UK standards (*Modernising Construction*, *Charter Handbook* and *Better Public Buildings*, see Chapter 3) or fully reflect the seven universal principles of supply chain management, especially the primary principle of lean thinking.

The Construction Industry Training Board (CITB) in the UK is liaising with various organisations to develop appropriate training workshops and courses, a good example of this is the ICOM/CITB link-up which offers a Diploma course in Construction Process Management that is built around the six goals of construction procurement best practice from the Construction Best Practice Programme booklet *A Guide to Best Practice in Construction Procurement* (see Further Reading for details). The CITB is also working with ICOM to develop layered action learning packages that start with operatives and end with senior managers, are based on the six goals of construction best practice and are appropriate for small and medium-sized enterprises and their supply chains, as well as larger companies.

Experience has shown that the inertia of deeply embedded traditional working practices could well be powerful enough to neutralise the change process, no matter how apparently beneficial it appears to the Chief Executive. Consequently, the Chief Executive will need to appoint a team both to develop the new working practices and then to work with the staff and operatives to ensure the application of those new working practices throughout the organisation. To ensure maximum effectiveness, the team will almost certainly need to work directly with the Chief Executive so that everyone in the organisation is left in no doubt about the Chief Executive's intentions and there is no possibility of the team being side-lined by those senior staff that are covertly against change.

As was made clear earlier in the chapter, the provision of appropriate training and coaching in their appreciation and application is of paramount importance to the successful adoption of new working practices. It is unreasonable to assume that people who have always worked in one way can suddenly work in a completely new way without a great deal of help and support. This help must take the form of mapping out in detail the changes to working practice and providing trainers who fully understand and are able to communicate the new processes.

In order to force the pace of change and test if everyone is applying the theory (rather than just using the appropriate buzzwords) the Chief Executive must ensure there is some way of measuring the improvements in performance that ought to be delivered by the new working practices. The result of the regular measurement of improved performance needs to be communicated in simple, straightforward terms to everyone in the organisation at reasonably frequent intervals. Ideally, this should 'name and shame' as well as giving public recognition to those that have been most successful in applying the new working practices and improving their performance.

It has always been said that 'success breeds success' and this is equally true of organisational change. As long as those involved in operating the changed working practices can see real and tangible improvements in the products those new working practices are delivering to their customers (and these can be internal as well as external customers) they will be motivated to continue to do things in this way. Consequently, there will be far less chance of the inertia of traditional working practices dragging them back to the old and inefficient ways of working.

Case History – Building Down Barriers

Whilst the morale of those at the sharp end will be boosted when they are an intrinsic part of the change process that delivers improved products to the client, the opposite is also true and their morale can be devastated if the changed working practices are not sustained. This was explained to me by the Chief Executive of the steel fabrication firm on one of the two Building Down Barriers pilot projects shortly after site completion, when he was trying to rebuild the morale of his firm following a bad experience on a warehouse project.

He said that at the completion of their Building Down Barriers project the morale of the fabricators and erectors had been sky high. Their skill and experience had been harnessed from the outset of the design of the steel frame and they had therefore been instrumental in achieving 'right first time' in the fabrication and erection of the steel frame.

The erectors had been able to get the architect and the structural engineer to understand the erection problems that were repeatedly encountered and they were convinced that their input at the start of design development was one of main reasons why the steel frame had been erected 'right first time' for the first time in their experience. The fabricators had also been able to get the architect and the structural engineer to understand the fabrication problems that were encountered time and time again and they were also convinced that their input at the beginning of design development was another of the main reasons why the steel frame had been fabricated and erected 'right first time' for the first time in their experience.

Because the fabricators and erectors had been fully in-volved in design development from the outset, and they had been respected and listened to by the architect and the engin-eers and their advice accepted, they felt valued in a way they had never felt before. They felt that the benefits of supply chain management were so obvious that no-one would want to go back to doing things in the usual fragmented and adversarial ways, especially since they could see ways of improving what they had achieved on the pilot project if they were able to work the same way on a subsequent project.

Unfortunately, the subsequent project was a conventionally procured warehouse and their skill and experience was not utilised by the consultant designers despite intense lobbying by their Chief Executive and warnings from him that the design was flawed. The outcome was a steel frame where the design was riddled with the usual errors and the fabricators knew that parts of the steel frame they were making would have to be taken down after erection and come back to be modified. The erectors also knew that they would have to dismantle parts of the frame after putting it up, take them back to the factory for modification, and then return to site to re-erect them.

Needless to say, the Chief Executive of the steel fabrication firm said the morale of his people had been lifted to unprecedented heights by their Building Down Barriers experience and had then been devastated by their subsequent warehouse experience. Yet the Chief Executive said that in reality the warehouse job was quite normal and the flaws in the design of the steel frame were no worse than usual.

The problem that neither I nor the Building Down Barriers development team had foreseen was that the erectors and fabricators would assume that what happened on the Building Down Barriers project was so sensible and beneficial it would be automatically replicated on all subsequent projects for all other clients and they would never have to go back to the old ways of working. They believed they would henceforth always be respected and valued members of both the design team and the construction team.

When reality struck on the next project and they had to go back to the adversarial and fragmented way they were always forced to work, they felt ignored and humiliated. They also felt that the Building Down Barriers project had given them false hope and had led them to believe in a new way of working that was not going to be possible against the antagonism and bloody-mindedness of the construction industry.

It is obvious from the EFQM and case history evidence that what the Chief Executive says and does is key to the success of any change process. The Chief Executive must be seen by every member of the work force to be totally committed to the changes in working practices and must be seen by both the work force and the customers to be determined that the new way of working should be the way the organisation operates for all its customers. The Chief Executive must be prepared to explain the ultimate goal for the change programme in a simple and easy to understand language and must also be prepared to ensure appropriate training is available and that all key members of staff attend it. Finally, he or she must be prepared to ensure that flow charts and explanations for the new working practices are developed and made available to staff who need them, and that the ensuing improvements in performance are regularly measured and the results published to the entire workforce. Unless the Chief Executive is prepared to lead the change process in this way, there is little point in even thinking about introducing a radical change in the way the organisation operates.

This all-important leadership role of the Chief Executive applies to every firm or organisation in the design and construction supply chain. It applies as much to the end user as it does to the manufacturer, since the application of the two key differentiators, the six primary goals of construction best practice and the seven universal principles of supply chain management (especially the primary principle of lean thinking) require radical changes in the way all of them operate.

The previous chapter explains the operation of the virtual company that is created by the long-term, strategic supply-side partnerships that are an essential part of supply chain management. It is imperative that every Chief Executive of every firm within the virtual company has a shared understanding of how the working practices need to be changed, why they need to be changed and how the improvements will be measured. It is also imperative that they are all equally committed to making the changes happen and that they are willing to be open with each other about what changes are going well in their firm and what is not going so well.

This all-important leadership role also applies to the vast number of institutes and trades organisations within the UK construction industry. They too must become an intrinsic, co-ordinated and supportive part of the change process if construction best practice and the seven universal principles of supply chain management (especially the primary principle of lean thinking) are to take hold and flourish. Their Chief Executives must provide co-ordinated leadership for the new, integrated way of designing and constructing that delivers the six primary goals of best practice. Unless they do so, the habitual fragmentation and confusion will continue, where each institute and each trades organisation appears to be heading in a different direction with a different view of what constitutes best practice.

This need for leadership from the pan-industry organisations given the task of leading the drive for radical reform was effectively grasped and demonstrated in the UK by Rethinking Construction in the autumn of 2002 when they published guidelines for public sector clients and based their six themes of construction best practice very directly and very precisely on the six goals of construction procurement best practice from the CBPP booklet *A Guide to Best Practice in Construction Procurement*.

This use in the UK by Rethinking Construction of the six CBPP goals to give a precise universal definition of what constitutes best practice sets an excellent example for the industry as a whole and should be imitated. The six goals from the CBPP booklet became the Rethinking Construction 'six primary themes of construction best practice' and are as follows.

'Six primary themes of construction best practice

☐ *The finished building will deliver maximum functionality and delight the end users.*

❑ *End users will benefit from the lowest optimum cost of ownership.*
❑ *Inefficiency and waste in the use of labour and materials will be eliminated.*
❑ *Specialist suppliers will be involved from the outset to ensure integration and buildability.*
❑ *Design and construction will be through a single point of contact.*
❑ *Performance improvement will be targeted and measurement processes put in place.'*

The issue of effective leadership is not confined to the construction industry: it must also include the further education establishments that provide courses in the various aspects of design and construction. Chief Executives or Heads of further education establishments must also play their part in supporting the change process. They must ensure their vision for the culture, structure and working practices of the industry matches that of the six goals of construction best practice and the seven universal principles of supply chain management, especially the primary principle of lean thinking.

They must ensure the graduates they produce share a common vision of the reformed industry that fits in with the direction those at the leading edge of reform have taken and accords with the true expectations of the end-user clients. They must abandon any assumptions they have that might continue to reinforce the traditional fragmentation and adversarial attitudes of the industry and they must actively reinforce the direction and the pace of reform towards:

❑ The total integration of design and construction necessitated by the seven universal principles of supply chain management.
❑ The elimination of all unnecessary costs in the design and construction process demanded by the primary, lean thinking principle of supply chain management.

❑ The delivery of the best whole-life value and best whole-life performance demanded by virtually all end users across the developed world.

Most importantly, those at the head of industry firms, industry institutes and federations, further education establishments, key pan-industry bodies such as Rethinking Construction or Be, and those in key positions at the DTI, must openly acknowledge that the first task of any leader is to establish precisely where they are. Unless they can be certain about current levels of performance, they cannot possibly set meaningful improvement targets.

This also requires leaders to agree on which aspects of performance have the greatest potential to reduce the capital cost of construction. As labour and materials account for around 80–90% of the cost of construction, and as the vast majority of this relates to work carried out by specialist suppliers (sub-contractors, trades contractors and manufacturers), it is obvious that the measurement of the effective utilisation of labour and materials by specialist suppliers should be given the highest priority.

Not only do construction contractors need to give top priority to measuring the effective utilisation of labour and materials by their specialist suppliers, they also need to give high priority to the development of systems that ensure that the improvements in the utilisation of labour and materials are recorded and that the resultant savings are captured and shared with the entire design and construction supply chain, including the end-user client, in the form of reduced tender prices for subsequent contracts and a reduction in the final account of the current contract.

If we go back to the four prime EFQM Excellence Model criteria listed at the start of this chapter, we can derive a series of top priority actions against each criterion, namely:

❑ **Leadership.** The leaders cannot develop viable improvement goals unless they know precisely where they and their suppliers are at present in terms of every aspect of performance at project and strategy level. Unless they first measure the effective utilisation of labour and materials throughout the design and construction supply chain, they cannot know how much improvement is either possible or desirable and they cannot prioritise the key areas for improvement. In fact, without performance measurement at project and strategy level, leaders could well waste a great deal of money and resources trying to improve aspects of performance that are already effectively done.

❏ **Processes.** An organisation cannot improve the management and execution of the working practices of its own people (and the people within its suppliers) unless it first carefully maps the new working practices, communicates them in a simple and easy to understand format to all the people who will have to operate the new working practices, and ensures that adequate and appropriate training exists to assist everyone at the sharp end to understand what they are to do differently.

❏ **Customer Results.** The firms in the design and construction supply chain should not merely assume that their constructed products are fully satisfying their end-user customers, especially in a situation where numerous end-user inspired reports are expressing considerable dissatisfaction. As in other sectors, customer satisfaction should be independently measured and the aspects that should be assessed with the greatest care are those that are of greatest importance to the competitive performance of the end user. The National Audit Office report *Modernising Construction* and the Confederation of Construction Clients, *Charter Handbook* provide excellent pointers to what is really important to end users in the UK, as do the corresponding reports in other developed countries such as the USA, Canada and Australia. The US report *National Construction Goals* was particularly effective at setting out the competitive needs of end users and explaining the ramifications for the end users of the poor performance of the construction industry.

❏ **Key Performance Results.** An organisation can only know what performance it is really achieving if it measures every aspect of its performance. In the case of construction contractors, since 80–90% of the total cost of construction comes from the labour and materials provided by the specialist suppliers (sub-contractors, trades contractors and manufacturers), there is little point in only measuring the performance of the construction contractor's own people. The full picture is only possible if the construction contractor ensures the specialist suppliers are measuring their effective utilisation of labour and materials and are making the information freely available to the construction contractor. Similarly, project managers, design consultants and quantity surveyors cannot give end-user clients any assurance about the elimination of unnecessary costs at project level, unless they have evidence derived directly from the measurement of the effective utilisation of labour and materials by every firm in the design and construction supply chain.

As the EFQM Excellence Model has not been widely taken up within the construction industry, the Construction Group within the British Quality Foundation has been working with an organisation called BQC Performance Management Ltd and the Construction Best Practice Programme to adapt the EFQM Excellence Model criteria for easier usage within the construction industry. The resulting guidebook is entitled *The Construction Performance Driver – A Health Check on your Business* and provides an excellent tool for improvement. In its section on leadership, under the question 'Am I a good boss?', it states:

'You are the one who decides most things in your company. Not only what gets done, but how. Your employees see you

around all the time and your "fingers are in all pies". It is likely that if anyone is going to pick up habits, both good and bad, that it will be from you. Have you realised yet that the way you manage your people and projects has a significant impact on whether your business is a success or not?'

The leadership section then goes on to pose a series of questions that the Chief Executive (and senior managers) need to ask themselves, such as:

❑ *'1.0 Are you always looking for improved ways of doing things?*
❑ *1.1 Do your management systems and methods encourage continuous improvement?*
❑ *1.2 Do you have effective ways to keep your employees informed of your desires and intent for the business?*
❑ *1.5 Do you have a plan to ensure you stay close to, and understand, your key customers and partners?*
❑ *1.6 Do you cultivate partnership-style relationships with your customers and suppliers?*
❑ *1.7 To avoid assumptions and misunderstandings, is there some formality in customer and supplier relationships?*
❑ *1.14 Do you ensure that resources are made available to support business improvement priorities?'*

The Construction Performance Driver – A Health Check on your Business is a must for any construction industry firm that is intent on improving its competitive edge and its profit margins. It is closely aligned with the radical reforms demanded by every report from Latham onwards and people at the sharp end of the construction industry (namely those that form the BQF Construction Group) have specifically devised its language to ensure ease of usage. Its statements under each section heading are remarkably perceptive and apt, for instance, under the question 'How do we do things?', it states:

'Ideally, you take your time, money, materials, ideas and ex-pertise and put together the things your customers value. The way you do this is crucial. You may have developed your methods bit by bit over the years, never having the time to stand back and see it all in perspective. If you have a few big customers you no doubt spend a lot of energy just trying to keep them happy but perhaps not looking beyond the fulfil-ment of the next contract or project. There are many different ways of doing things but whatever your approach, it is continu-ally working at improving the way you do things that could well hold the key to beating the competition or, in some cases, surviving at all.'

It follows from the above that effective leadership demands a deep and clear understanding of what needs to be done differently, why it needs to be done differently, what improve-ments in performance will result from doing things differently, and how both the current performance and the performance resulting from those improvements will be measured.

Most importantly, effective leadership in the formation of the long-term, strategic supply-side partnerships that are an essential first step in supply chain management, lean think-ing and the formation of a virtual company demands the use of clear and unambiguous terminology. The Chief Executive and the senior management team cannot afford to use vague buzzwords without defining them in terms everyone can understand.

Always remember the excellent and well founded advice given by the CEO of the aerospace company I described at the start of this chapter. Over three years he took 30% off the cost of production of the aero-engines his company produced and reduced production time by 50% without any reduction in the outstanding whole-life quality or whole-life performance of the engines. On the basis of the evidence of this massive improvement in performance, his advice ought to be fully heeded by CEOs and senior man-agers in the construction sector.

A radical change in well established working practices will only happen if everyone, down to the man that makes the tea on site, understands exactly what has to be done differently, why it has to be done in the new way, who has to do what in a new way, and what aspects of performance will be measured to show the improvements have reduced the underlying labour and materials costs. The following case history from the Building Down Barriers project illustrates this point very effectively.

Case History – Building Down Barriers pilot project

On one of the two Building Down Barriers pilot projects, the main contractor's young project manager was particularly effective at explaining in plain English what the application of the seven universal principles of supply chain management meant to everyone involved in the full supply-side design and construction team.

This was demonstrated by the man on the gate into the site who became a knowledgeable and enthusiastic champion of supply chain management. I well recall him stopping my Chief Executive at the gate and giving him an excellent induction into how the site operated, why it was different from any other site and what his contribution was to getting it right first time. He explained clearly and enthusiastically:

❑   what his role was on the site
❑   how it differed from what he did on all previous sites
❑   why it made sense for him to operate in this new way
❑   how his new way of working helped the tradesmen to achieve 'right first time'
❑   why this new way of working made the site a far happier place to work
❑   why it would make sense for all sites to work in this new way
❑   how much he enjoyed working in this new way

He strongly emphasised that the way of working on this site was totally different from any other site on which he had worked in his long career in the building industry. He said that on this site the overriding aim was to make sure that the tradesmen were able to work without disruption or delay. He said that he had been encouraged to talk directly to the various tradesmen about the things that disrupted or delayed their work and between them they had agreed that there were things that he could do to help, namely:

❑ At the end of the working day, he went around the site and tidied away materials and any debris so that there was nothing to impede or disrupt the work of the individual tradesmen the next day.

❑ He constantly liaised with the tradesmen to find out where they would be working over the next few days and thus where they needed the next delivery of materials to be placed. He also made sure he knew which areas of the site had to be kept clear of materials deliveries to avoid disrupting work.

❑ He stopped every delivery vehicle at the gate, checked what it carried and then directed it to unload its materials where he knew the relevant tradesmen needed them.

❑ He stopped all other vehicles at the gate and directed them to park in a position where they would not disrupt work or materials delivery.

❑ He explained to every new arrival at the site gate (as he had done with my Chief Executive) why and how the site was different, what his job involved and why the new way they were all working made sense and made it a far more efficient site.

❑ Finally, he said that this was the first site he had ever worked on where the tradesmen were happy in their work and this was because they were never made cross because something or someone was in the way of what they wanted to do.

Whilst he did not use the usual buzzwords in his explanation of what his role was on the pilot project and what made the pilot project totally different, his understanding of what supply chain management meant to the people on site was exemplary. He obviously felt properly valued and respected for the first time and he was obviously delighted to have been given the freedom to make a major contribution to getting it right first time on site.

His most powerful comment to my Chief Executive was that everyone enjoyed working on the site because everyone was part of the team, everyone was listened to and their advice heeded, and everyone could get on with their job without disruption. At the end of our visit to the project, my Chief Executive turned to me and said that the gateman had been right, the site truly was a happy site and it was the first time he had ever come across such a situation.

The above case history also illustrates the need for people in leadership positions to use simple and straightforward language when they are trying to get their message across. Because the young project manager understood what supply chain management meant, he was able to avoid all the meaningless buzzwords and he used words that could be properly understood by the people on site when cascading the message down to his design and construction team.

In the next chapter, I endeavour to ease the communications process by offering explanations and definitions (often drawn from other business sectors) that cut through the usual confusion over the precise meaning of the many and various buzzwords that are in constant use around the construction industry when performance improvement is discussed. At present, it seems to me that because the buzzwords are not defined or explained in simple terms, they often form a major barrier to improvement because of the confusion and misconceptions they create.

9 The Buzzwords Explained

Unlike other business sectors, where there are several decades of experience of performance improvement through the use of supply chain management and lean thinking tools and techniques, the lack of experience of supply chain management within the construction industry means that there is considerable confusion over the exact meaning of the many buzzwords that are bandied about when those in the industry are talking about improving performance. As I said in Chapter 1, this was picked up effectively in a leading article in the UK *Building* magazine in March 1999 by Andrew Sims, who said:

'The most famous buzzword of all, partnering, has been subject to a lot of abuse. It has been hijacked by consultants and corrupted by contractors. Many firms have been guilty of cherry-picking the bits that suited them and discarding the rest.'

In this chapter, I endeavour to explain in simple terms each of the buzzwords that are in common use within the construction industry. I have restricted my explanations to accord with the normal usage of such terms across other business sectors, as it seems wise to use terminology that is common to the rest of the business world. In some cases this will mean that the explanation appears to conflict with what is being assumed by many in the construction industry, but if the industry is to do as advocated by the various reports on improvement and learn from the experiences of other business sectors, it obviously needs to start using the same language as those other sectors.

The buzzword definitions that appear in this chapter are drawn from sources that are well established in other business sectors, such as the universally renowned European Foundation for Quality Management (EFQM) Excellence Model and the SCRIA (Supply Chain Relationships in Aerospace) handbook *Working Together*, that was sponsored by the UK Department of Trade and Industry, with supporting explanations where necessary drawn from my experiences on the Building Down Barriers project and the radical improvement programme at Defence Estates.

The buzzwords and their explanations are listed in the following sections.

Benchmark

The EFQM definition is as follows:

'A measured, "best-in-class" achievement; a reference or measurement standard for comparison; this performance level is recognised as the standard of excellence for a specific business process.'

In the case of Building Down Barriers, we recognised that effective supply chain management did not exist at all within the construction industry. If we wished to develop a set of process-based supply chain management tools that construction industry firms could use to drive out unnecessary cost and drive up whole-life quality, we needed to look beyond the construction industry for best practice in supply chain management. As we had no idea where to look for best practice, I approached the Warwick Manufacturing

Group for advice because they had an international reputation for expertise in supply chain management. I asked them if they could point us to a best-in-class manufacturer that we could use as a benchmark for best practice in supply chain management. We then appointed the Warwick Manufacturing Group to work with us to convert the best-in-class manufacturer's supply chain management process into a set of tools that would enable construction industry firms to import that benchmark supply chain management process.

In the case of Defence Estates, we used the EFQM to locate examples of best practice in the different aspects of our business. In my own case, I discovered that TNT had radically changed the way it used its operations manuals because its EFQM Excellence Model self-assessments had exposed how ineffective and unsuitable the manuals were at the sharp end of their business. Subsequent to the change, the manuals were replaced with simple flow charts that were hung on walls adjacent to every operation and only the training staff had access to the manuals, which were used to teach the operatives how to understand and use the flow charts. The effectiveness of this improvement had been validated in the subsequent EFQM self-assessments.

Since my area of responsibility at Defence Estates included the production of guides and manuals, I immediately recognised the sense of what TNT had done and also recognised that the lesson was equally applicable to Defence Estates' field of operation; in short, TNT provided me with an excellent best-in-class standard that I could import into Defence Estates in order to make our guides and manuals more effective at the sharp end.

Benchmarking

The EFQM definition is as follows:

'A systematic and continuous measurement process; a process of continuously comparing and measuring an organisation's

business processes against business leaders anywhere in the world to gain information that will help the organisation take action to improve its performance.'

In the case of Defence Estates, we needed to compare our rate of improvement with other organisations, both within the public sector and elsewhere, to see if we were improving as quickly as others, especially other central government departments. The EFQM Excellence Model provides a unified structure to self-assessment and a unified marking system that ensures that the overall self-assessment mark from all kinds of businesses can be compared. Thus we were able to compare our rate of improvement with others and we could be sure that trying to compare the results coming out of totally different measurement systems was not misleading.

Client

There is no EFQM, SCRIA or other business sector definition of the term 'client', since such a concept does not exist in other business sectors. There the term 'customer' is used to define the person or organisation on the demand-side to which a product, good or service is being sold. In the construction industry, the term 'client' covers the multiplicity of organisations and people on the demand-side, such as property divisions, developers, the end users of the building, and even occasionally the in-house architects and engineers.

Construction best practice

This is defined in the UK Rethinking Construction publication *Rethinking the Construction Client – Guidelines for Construction Clients in the Public Sector* as the following:

❑ *'The finished building will deliver maximum functionality and delight the end user.*

❑ *End users will benefit from the lowest optimum cost of ownership.*
❑ *Inefficiency and waste in the use of labour and materials will be eliminated.*
❑ *Specialist suppliers will be involved from the outset to ensure integration and buildability.*
❑ *Design and construction will be through a single point of contact.*
❑ *Performance improvement will be targeted and measurement processes put in place.'*

These six goals of construction best practice were taken directly from the UK Construction Best Practice Programme booklet *A Guide to Best Practice in Construction Procurement* where they were deduced from a detailed analysis of the various UK and other country's reports that had demanded radically improved performance from the construction sector.

It is interesting to note that the six goals could apply equally well to manufacturing best practice, which of course means that they pass the test imposed in every report in every country, namely, that the construction industry should import best practice from other business sectors.

Critical Success Factors (CSFs)

The EFQM definition is as follows:

'The prior conditions that must be fulfilled in order that an intended strategic goal can be achieved.'

In the case of Building Down Barriers, if we wanted to be sure that our supply chain management toolset really worked, we had to test and refine the tools on live projects. Consequently, one of our Critical Success Factors (or one of the things that we had to have in position, or have available, before we could achieve our goal of developing a viable and proven supply chain management toolset) was to persuade the Army to give us two similar building projects and to guarantee that both would proceed at the same time so that they could be used as test-bed pilot projects for the developing toolset.

Another CSF was the need to be certain that everyone involved, including the two pilot project design and construction teams, had (and continued to have) a common understanding of what the Building Down Barriers project was about so that we were all heading in the same direction and at the same speed. This CSF was checked at six-monthly intervals by holding workshops for the entire team at which the Tavistock Institute (the internationally acknowledged experts on cultural change within organisations that were the project managers of the Building Down Barriers project) and Warwick Manufacturing Group (the internationally acknowledged experts on supply chain management within the Building Down Barriers team) carefully assessed everyone's level of understanding and then ensured any gaps were plugged before the workshop closed.

Yet another CSF was the rate of development of the supply chain management tools. Until the Tavistock Institute and Warwick Manufacturing Group development team had produced a draft tool, the two pilot projects could not move forward. In fact, they could not even start producing the two project briefs until the tool that related to the use of value management was available because we wanted to ensure that the end users and the specialist suppliers were fully involved in a specific way. Interestingly, the problems the two pilot project teams had in understanding and applying the draft supply chain management tools generated an additional CSF, because the development team had to urgently refine the tools in close collaboration with the two pilot project teams in order to avoid design development on the pilot projects coming to a halt.

In the case of Defence Estates, the Chief Executive and the senior managers held a series of workshops that were chaired by an external expert and at which we hammered out the things we had to have available at each stage of the improvement process if we were going to be able to achieve our goals. One of these was the development of a networked user-friendly business management system (BMS) that was loaded with all our new business processes set out as very simple flow charts, each of which could be located and recovered within seconds. We recognised that our improved way of working would not happen until the BMS was in position and could provide our widely dispersed staff with the information they needed to enable them to start changing their outdated working practices to the new working practices, and to do so in a uniform way across the entire organisation.

Defects

As defects are relatively uncommon in the manufacturing sector (you would never want to fly in an aeroplane that was delivered 'practically complete' with a long list of manufacturing defects that would be corrected during the first six months of operational flying), there is neither an EFQM nor a SCRIA definition of what constitutes a defect. However, the UK *Clients' Charter Handbook* describes a defect as anything that failed to perform to the demand-side customer's satisfaction (which included the end users) during the whole life of the construction solution. This included functional performance as well as the performance of components and materials.

In the case of the two Building Down Barriers pilot projects, a defect was defined as a deviation from any aspect of the 35-year whole-life cost prediction, including the predicted functional performance of the building. If a component or material failed prematurely, if the maintenance or the running costs (especially fuel or energy costs) proved to be significantly higher that predicted, then the deviation would be deemed a defect and there would be an expectation that the supply-side design and construction team (the prime contract team) would rectify it.

End user

The end users are the people that will occupy and use the building or facility when it is constructed. The many reports on improving the performance of the construction industry, particularly the US *National Construction Goals* report and the UK *Rethinking Construction* report, emphasise the primary importance of judging the success of a building by measuring the delight of the end users. In other business sectors, the more accurate term 'end customer' is used to define the ultimate user of a product or good.

The concept of an end user appears to have developed in the construction industry because of its uniquely fragmented nature, which has led to parts of the industry (such as the professional design team members) becoming part of the demand side. The people who occupy and use the building or facility can be separated from the supply-side construction team (especially the specialist suppliers) by a multitude of layers. These can form communication barriers that block the detailed functional and business needs of the end users being fully understood by the people who construct the building. The various reports from across the developed world that are critical of the poor performance of the construction sector make it clear that the capability of a building to meet the users' needs over its useful life is of primary importance to the competitive success of the end user.

As a consequence, satisfying the end user of any building or facility is by far and away the most important goal for the design and construction team. Thus, the end user is the most important part of the whole demand-side team and should not be lost sight of under the all-encompassing term 'client'.

Improvement/performance improvement

The Warwick Manufacturing Group at Warwick University provided the Building Down Barriers team with the following definition of performance improvement from their experience of best practice in both the manufacturing and retail sectors.

'Real improvement in performance comes from the elimination of unnecessary costs (in the form of the inefficient utilisation of labour and materials) from the underlying labour and materials costs that make up the bulk of the total of the total cost of construction.'

In other business sectors, there is a clear belief that since at least 80% of the total cost of any good is made up of the underlying labour and materials costs, any real improvement in performance must come from the elimination of any unnecessary underlying labour and materials costs. Thus the term 'improvement' always relates to the continuous search for all forms of unnecessary costs in the effective utilisation of labour and materials. This need to seek out and remove wasted time and effort across all the firms in the design and construction supply chain is also explained under the explanation for 'Lean Thinking'.

Integration/integrated working

As in other business sectors, this relates to the total integration of the entire supply-side design and construction supply chain in the same way that a best practice manufacturer, such as Toyota, would fully integrate their design and manufacturing supply chain to form a virtual company. The purpose is to facilitate lean thinking across firms that make up the total supply chain and thus eliminate all forms of unnecessary costs in the underlying labour and materials costs. The *Building Down Barriers Handbook of Supply Chain Management* (see Further Reading) talked of supply chain integration and of integrated working of the supply chain in an effort to avoid confusion about the correct meaning of integration and integrated working.

As in other business sectors, the demand-side client and end user should not be an intrinsic part of the integrated supply-side design and construction partnerships that make up the virtual company. The responsibility of the demand-side team is only to do business with a supply-side design and construction virtual company that can prove that all its constituent firms are already working together in a fully integrated and collaborative way that utilises supply chain management and lean thinking tools and techniques. Having contracted with a virtual company, the demand-side client should then ensure that the end users interface directly with all members of the supply-side design and construction team (especially the specialist suppliers or trades contractors) at all stages of the design and construc-tion process in order to ensure the people designing and constructing the building are fully aware of the detailed functional and business needs of the end users.

Key Performance Results

The EFQM definition is as follows:

'Those results that it is imperative for the organisation to achieve.'

In the case of Building Down Barriers, the Key Perform-ance Results related entirely to the deliverables on the two pilot projects. Both had to be completed, be free of defects and ready for operation by the target deadline set by the end users. Both had to provide the end users with an accurate prediction of the cost of ownership over the 35-year life of the buildings and the predicted cost of ownership (whole-life cost) had to be significantly below the end user's estimated cost of ownership when both were converted into Net Pre-sent Value figures (the amount of capital that would have to be set aside and invested to cover the cost of ownership over the 35-year life of the buildings). Both had to deliver all the functional requirements listed and described in the project briefs and do so in a way that ensured maximum functional efficiency, and this had to be confirmed by the end users once they had taken up occupation and started using the buildings.

In the case of Defence Estates, the first Key Performance Result was the awarding of the first prime contract. Until a totally integrated design, construction and maintenance team had been awarded the first prime contract, we could not claim to have achieved what we set out to do.

Lagging indicators

The EFQM definition is as follows:

'Lagging indicators show the final outcome of an action, usually well after it has been completed. Profitability is a lagging indicator of sales and expenses. Perception measures are also referred to as lagging (trailing/following) indicators. A perception result relates to direct feedback from a stakeholder, e.g. when employees respond via an internal attitude survey.'

By this definition, many of the Construction Best Practice Programme Key Performance Indicators are lagging indicators. It could be argued that they would be far more effective in driving forward the kind of improvements in performance demanded by end users if the Construction Best Practice Programme Key Performance Indicators had all been the leading indicators I have suggested below, instead of lagging indicators.

In the case of Building Down Barriers and its two pilot projects, there were no lagging indicators used because we needed to know what was happening as and when it happened, not a year later, by which time it was far too late to take action to change and improve the development of the various tools or improve the effectiveness of specific design and construction activities. We recognised that it was imperative that we had prompt feedback from the pilot projects to the toolset development team so that the development team could quickly see what refinements to existing tools were needed, or what additional tools were required to keep the whole exercise on track.

In any drive to improve performance, lagging indicators are of dubious benefit because the information they provide always comes far too late to allow those at the sharp end of the improvement process to modify what they are doing where things are not improving as intended.

Leading indicators

The EFQM definition is as follows:

'Leading indicators, sometimes referred to as driving indicators, are usually measured more frequently than lagging indicators. They are the result of a measurement process that is driven by the organisation itself and is entirely within their span of control, e.g. measuring process cycle times. Leading indicators are those that predict, with a degree of confidence, a future outcome. Employee satisfaction, although a lagging indicator for the morale of the staff, is usually recognised as a leading indicator of customer satisfaction.'

It follows from this definition that leading indicators in the construction industry's improvement drive ought to be those things that should be continuously improved at project level in order to deliver the better value demanded by end users. It therefore makes sense for UK construction industry firms to include in their leading indicators those improvements listed in the *Charter Handbook* (in the USA it would be sensible to use the National Construction Goals). The *Charter Handbook* improvements were described earlier in this book, but as a reminder they are listed as follows:

❏ major reductions in whole-life costs
❏ substantial improvements in functional efficiency
❏ a quality environment for end users
❏ reduced construction time
❏ improved predictability on budget and time

❑   reduced defects on handover and during use
❑   elimination of inefficiency and waste in the design and
construction process.

As the *Charter Handbook* gives the above improvements
top priority, it would be sensible for project teams to set
themselves leading indicators (I have called them Key
Performance Indicators in this book to accord with the
approach adopted by the Construction Best Practice Programme) that directly relate to each of the seven *Charter
Handbook* priority improvements in performance. Since
the elimination of inefficiency and waste dates back to the
1994 Latham report, it would make sense for project teams
to give top priority to measuring the efficient utilisation of
labour and materials and set themselves a target of reducing
it by X% per project as a leading indicator.
   At a strategic level, construction industry firms ought to
set leading indicators across all their projects that relate to
the same seven priority areas for improvement so that their
leading indicators at all levels are co-ordinated, i.e. everyone
in their firm is going in the same direction. Needless to say,
the same leading indicators ought to be used by their suppliers (see the EFQM definition of suppliers under 'Supply
Chain' below) or chaos will reign.

Lean thinking

The UK *Rethinking Construction* report in 1998 described
lean thinking as follows:

*'Lean Production is the generic version of the Toyota Produc-
tion System, recognised as the most efficient production
system in the world today. Lean Thinking describes the core
principles underlying this system that can also be applied to
every other business activity – from designing new products
and working with suppliers to processing orders from customers. Few products or services are provided by one organisation
alone, so that waste removal has to be pursued throughout the*

whole value stream – the entire set of activities across all the firms involved in jointly developing the product or service. New relationships are required to eliminate inter-firm wastage and to manage the value stream as a whole. Removing wasted time and effort represents the biggest opportunity for performance improvement. Lean Thinking represents a path of sustained performance improvement and not a one-off programme.'

Thus lean thinking is the endless search for, and elimination of, all forms of unnecessary cost in the underlying labour and materials costs, which is also the primary principle of supply chain management in *Building Down Barriers Handbook of Supply Chain Management,* namely: *'Compete through superior underlying value'.* The handbook goes on to describe this as follows:

'The commercial core of supply chain integration is setting up long-term relationships based on improving the value of what the supply chain delivers, improving quality and reducing underlying costs through taking out waste and inefficiency.'

The handbook then goes on to point out that:

'This is the opposite of "business as usual" in the construction sector, where people do things on project after project in the same old inefficient ways, forcing each other to give up profits and overhead recovery in order to deliver at what seems to be the market price. What results is a fight over who keeps any of the meagre margins that result from each project, or attempts to recoup "negative margins" through "claims". The last thing that receives time or energy in this desperate, project-by-project, gladiatorial battle for survival is consideration of how to reduce underlying costs or improve quality.'

Partnering

The UK Department of Trade and Industry sponsored Sigma Management Development Ltd SCRIA (Supply

Chain Relationships in Aerospace) handbook *Working Together* defines a supply chain as follows:

'The enlightened companies have recognised that for the supply chain to work to its optimum, the flow of information has to be excellent. They have selected a sub-set of suppliers with whom they form closer relationships in order to facilitate the information flow. These closer relationships can be regarded as a form of partnership. The key to "oiling the wheels" of the supply chain is for companies to decide which suppliers have the potential to add most value to their business and agree a form of partnership. There are no pre-defined rules for a partnership. Each one should be constructed so as to be appropriate to the relationship required.'

The EFQM definition of partnering is as follows:

'A working relationship between two or more parties creating added value for the customer. Partners can include suppliers, distributors, joint ventures, and alliances. Note: Suppliers may not always be recognised as formal partners.'

Building Down Barriers Handbook of Supply Chain Management says of partnering:

'The commercial core of supply chain integration is setting up long-term relationships based on improving the value of what the supply chain delivers, improving quality and reducing underlying costs through taking out waste and inefficiency.'

It follows that long-term, strategic supply side partnering is an essential prerequisite of lean thinking.

As in all other business sectors, partnering ought to relate primarily to the formation of long-term, strategic supply-side relationships between the firms that make up a design and construction supply chain capable of delivering a comprehensive range of building types and construction activities for a variety of demand-side customers (small and occasional as well as major repeat customers). In the UK aerospace sector, the organisation that results from these long-term, strategic supply-side relationships is referred to as a 'virtual company'.

The purpose of such strategic supply-side partnering relationships is to enable the supply-side design and construction firms to work together at both project and strategy level to continuously drive out all forms of unnecessary costs (caused by the inefficient utilisation of labour and materials) and to continuously drive up whole-life quality.

The output of such strategic supply-side partnering should be the continuous conversion of unnecessary costs into lower prices and higher profits, whilst improving the whole-life value of the building or the constructed product.

Where a demand-side customer has a sufficiently large workflow of construction projects (including maintenance activities), there may be benefits to be gained from a partnering relationship with a construction industry firm. However, such a partnering relationship is secondary to the primary supply-side partnering relationships and should only be considered if the supply-side organisation can prove that it has well established, long-term, strategic partnerships in place with the key members of its supply-side design and construction supply chain.

Performance measurement

As in all other business sectors, it is important to focus performance measurement directly on those aspects of performance that contribute the greatest proportion of the overall cost of construction. This is particularly important if there is a wish to improve performance and thus reduce the cost of manufacturing or the cost of construction.

As mentioned before, the construction sector is remarkably similar to the manufacturing sector, in that over 80% of the total cost of construction is made up of the underlying labour and materials costs. Virtually all of these costs come from the specialist suppliers (design professionals, trades contractors, specialist contractors and manufacturers) since the main construction contractor's directly employed people rarely make up more than 5% of the total cost of construction. It follows from this that the construction sector ought to follow the best practice approach of the manufacturing sector and focus on accurately measuring its leading indicator of performance.

Other aspects of performance may also be measured, but most of these will be lagging indicators (such as profitability) that relate to the final outcome of a whole series of interrelated activities. These usually come at a stage when it is far too late to take action to improve performance on the project that created the profits (or losses) and generally tell you nothing about what caused the lagging indicator to be what it was measured to be.

Right first time

This is the ultimate goal of lean thinking, which is the primary principle of effective supply chain management. 'Right first time' is about the removal of all forms of inefficiency and waste in the design and construction process, such that every activity in the process is done 'right first time' without reworking, disruption, delay or interruption of any kind. This requires careful preplanning and the close involvement of trades contractors, specialist contractors and manufacturers in design development from the outset in order to inject their knowledge and experience of buildability into design at a stage when it can make a difference.

The architect cannot achieve a design that can be constructed 'right first time' if the key trades contractors, specialist contractors and manufacturers are not closely involved from the outset of design development and their advice on buildability is ignored. Similarly, the trades contractors and specialist contractors cannot construct their aspect of the construction right first time unless they have ensured that their extensive knowledge and experience of buildability is incorporated into the developing design and that they have collaborated with the other members of the design and construction team to ensure that their aspect of construction is carefully co-ordinated with all other aspects of construction with which it has an interface.

It therefore follows that 'right first time' is unlikely to be achieved without the existence of long-term, strategic supply-side partnerships that are essential to the creation of open, collaborative, supportive and trusting relationships between the firms that make up the design and construction supply chain.

Specialist supplier

This term covers every firm in the whole supply-side design and construction supply chain. An architectural firm is a specialist supplier of architectural services, a steel fabrication firm is a specialist supplier of steel frames, a project management firm is a specialist supplier of project management services and a scaffolding firm is a specialist supplier of scaffolding. The term merely refers to a firm that supplies a particular service, component or material. In other business sectors, the term 'specialist supplier' would rarely be used and the more straightforward and logical term 'supplier' would be used to describe any firm supplying services, components or materials within the overall supply chain.

Supplier

See the explanation for 'Specialist Supplier' above. In other business sectors, the term 'supplier' logically and simply covers any firm that supplies services, components and materials within the overall supply chain. Under this definition, a supplier can supply either manufactured (or constructed) products or services, thus everyone on the supply-side is defined by the EFQM Excellence Model as a supplier. This differs from the somewhat confused practice in the construction industry where construction contractors generally restrict the use of the term 'supplier' to mean someone who supplies a manufactured or constructed product. Those firms that supply services are referred to using a variety of terms, such as 'sub-contractor', 'trades contractor', 'specialist contractor' or 'consultant'.

Consequently, when someone from outside the construction industry uses the term 'supplier' and assumes the industry will understand it to mean what the EFQM means by the term, confusion can reign. In my view, the construction industry would be well advised to start using the same language as other sectors and thus use the term 'supplier' to mean all those within the design and construction supply chain that produce either products or services.

Supply chain

The EFQM definition is as follows:

'The integrated structure of activities that procure, produce and deliver products and services to customers.'

The UK Department of Trade and Industry-sponsored Sigma Management Development Ltd SCRIA Handbook *Working Together* defines a supply chain as follows:

'A supply chain exists to design, engineer and build products and services for the end customer. It is a relationship of many companies, because the end product is so complex that no single company could possibly build it. It can be helpful to regard the supply chain as if it were a virtual company which exists purely to satisfy the demands of the end customer.

What is clear from these two definitions is that in other business sectors the end customer or end user is not part of the supply chain, no matter how closely they might need to interface with the supply chain.

Supply chain management

Building Down Barriers Handbook of Supply Chain Management defined this as:

'*Replacing short-term single project relationships with long-term, multiple project relationships based on trust and co-operation. These standing supply chains focus on delivering value as defined by their clients. Long-term strategic supply chain alliances can incorporate continuous improvement targets to reduce costs and enhance quality, and focus on the through-life cost and functional performance of buildings. The idea of continuous improvement, based on a systematic analysis of the weaknesses and strengths in existing design and construction processes, underpins every aspect of the Building Down Barriers approach to supply chain management. Without this discipline, it would be impossible to reduce through-life costs significantly, or enhance quality, deliver superior functionality or any other design benefits, or improve levels and certainty of profits for the supply chain.*'

The *Building Down Barriers Handbook of Supply Chain Management* then went on to explain why effective supply chain management is governed by the following seven universal principles.

The primary principle of effective supply chain management:

❑ Compete through superior underlying value (i.e. compete through being more effective in your utilisation of labour and materials by eliminating all forms of unnecessary costs). This is the important lean thinking aspect of supply chain management.

The supporting principles of effective supply chain management:

❑ Define client values (i.e. use value management tools and techniques to find out precisely what the end user's functional and business needs are, both in general and in detail).

❑ Establish supplier relationships (i.e. set up long-term, strategic supply-side partnerships with all other key firms in the design and construction supply chain).

❑ Integrate project activities (i.e. set up cluster teams of all those involved in particular construction activities that have complex interfaces to ensure 'right first time' is achieved on site).

❑ Manage costs collaboratively (i.e. professional designers need to collaborate closely with the trades contractors, specialist contractors and manufacturers to drive out all forms of unnecessary costs in the effective utilisation of labour and materials).

❑ Develop continuous improvement. (This is at strategic level within the long-term, supply-side partnerships and is about eliminating all forms of unnecessary costs, driving down whole-life costs and driving up whole-life performance.)

❑ Mobilise and develop people (i.e. provide the people in the long-term, strategic supply-side partnerships with appropriate and adequate expert coaching, training and support).

Virtual company

The UK Department of Trade and Industry-sponsored Sigma Management Development Ltd SCRIA handbook *Working Together* says of a virtual company:

'*A supply chain exists to design, engineer and build products and services for the end customer. It is a relationship of many companies, because the end product is so complex that no single company could possibly build it. It can be helpful to regard the supply chain as if it were a virtual company which exists purely to satisfy the demands of the end customer. This* company is very complex with many "departments", each with its own capability to add value into the process of making and delivering the final product. As with any company, relationships between departments need to be excellent if they are to work together to achieve optimum results.

However, those companies that have realised that to achieve their corporate objectives, it is essential to have all departments aligned to the end goals, are now realising that the same has to happen within the virtual company in which they exist. Each "supplier" needs to be empowered to add value to each "customer" right through the chain. Two vital ingredients for successful empowerment are open communication and trust.

The enlightened companies have recognised that for the supply chain to work together to its optimum, the flow of information has to be excellent. They have selected a sub-set of suppliers with whom they form a closer relationship in order to facilitate the information flow. These relationships can be regarded as a form of partnership. The key to "oiling the wheels" of the supply chain is for companies to decide which suppliers and customers have the potential to add most value to their business and agree a form of partnership. There are no pre-defined rules for a partnership.'

Further Reading and Help

For a concise, plain-English explanation of best practice procurement:

A Guide to Best Practice in Construction Procurement

Available from the Construction Best Practice Programme, PO Box 147, Watford, WD25 9UZ. Tel: 0845 605 55 56 Email: helpdesk@cbpp.org.uk Website: www.cbpp. org.uk

A simple guide written by the author of this book and aimed at working-level staff in all sectors of the industry, from end users to manufacturers. It explains the historical background to the Rethinking Construction movement and briefly describes the key aspects of the three best practice standards (*Better Public Buildings, Charter Handbook* and *Modernising Construction*). It then uses the analysis to list the six goals of construction procurement best practice (these were subsequently adopted by the UK Rethinking Construction organisation as the six themes of construction best practice). It also sets out the next step actions that each sector must take in order to achieve the best practice of the three standards and warns of the consequences of doing nothing.

For a detailed, plain-English guide to best practice in design and construction:
Building Down Barriers – A Guide to Construction Best Practice
ISBN 0-415-28965-3
Available from Spon Press, 11 New Fetter Lane, London, EC4P 4EE or 29 West 35th Street, New York, NY 10001.

A comprehensive and detailed guide written by the author of this book and aimed at working-level staff in all sectors of the industry, from end users to manufacturers, and developed from the concise Construction Best Practice Programme booklet *A Guide to Best Practice in Construction Procurement*. This book explains and compares, simply and clearly, the main aspects of the three UK best practice standards (*Better Public Buildings*, *Charter Handbook* and *Modernising Construction*), compares them with similar developments in countries such as the USA and Singapore and lists the six goals of construction best practice that satisfies them all (these were adopted by the UK Rethinking Construction organisation as the six themes of construction best practice). It goes on to explain in detail the fundamental culture changes they necessitate in each sector of the UK construction industry and provides specific action plans for each sector that should deliver those cultural changes. The book is also a valuable guide for students in all design and construction courses.

For a detailed, plain-English guide to performance measurement in construction:
Performance Measurement for Construction Profitability
ISBN 1-4051-1462-2
Available from Blackwell Publishing Ltd, 9600 Garsington Road, Oxford, OX4 2DQ or from Blackwell Publishing Inc, 350 Main Street, Malden, MA 02148-5020, USA.

A comprehensive and detailed guide written by the author of this book and aimed at working-level staff in all sectors of the industry, from end users to manufacturers, developed from the concise Construction Best Practice Programme booklet *A Guide to Best Practice in Construction Procurement* and linked with the CBPP/BQF-sponsored publication *The Construction Performance Driver – A Health Check for your Business*. This book uses the same historic background material as *Building Down Barriers – A Guide to Construction Best Practice* but goes into far greater detail about performance measurement. The book provides a highly practical, easy to read guide, focusing strongly on the day-to-day needs of managers at all levels. It uses the everyday business language of construction firms to explain how to set up and run performance measurement, self-assessment and benchmarking systems. It is comprehensive and informative with plenty of real-life examples and, most importantly, tells you what to do differently on Monday.

For a simple guide to self-assessment using the EFQM Business Excellence Model:
The Construction Performance Driver – A Health Check for your Business
ISBN 1-90216-912-3
Available from BQC Performance Management Ltd., PO Box 175, Ipswich, IP2 8SW. Tel: 01473 409962 Fax: 01473 409966 Email: helpdesk@bqc-network.com Website: www. bqc-network.com
The guide provides an easy to use assessment tool, written specifically for the construction industry with considerable input from those at the sharp end within construction industry firms and based on the model of Business Excellence that has been widely used in major European and UK organisations since the early 1990s. The guide has been developed for use by all organisations operating within the construction industry (large or small, public or private) and who have a desire to improve their current business performance.

For the key UK reports advising on best practice in construction procurement:
Better Public Buildings
Available from the Department of Culture, Media and Sport, 2–4 Cockspur Street, London, SW1Y 5DH. Tel: 020 7211 6200 Website:www.culture.gov.uk/pdf /architecture.pdf
A short, lucid guide (six pages of text) that focuses strongly on the business benefits of well designed buildings that enhance the quality of life, and therefore the efficiency, of the end users. It also

explains the business benefits of using whole-life costs as the basis of design and construction decisions and it makes clear that best practice necessitates the appointment of integrated design and construction teams.

The Clients' Charter Handbook

Available from the Confederation of Construction Clients, 1st Floor, Maple House, 149 Tottenham Court Road, London, W1T 7NF. Tel: 020 7554 5340 Fax: 020 7554 5345 Email: cccreception@ccc-uk.co.uk
Website: www.clientsuccess.org

A short, lucid guide (12 pages of text) that explains the approach to construction procurement that every chartered client must adopt. It focuses strongly on the importance of the client's leadership role within an integrated design and construction supply chain, which targets major reductions in whole-life costs, substantial improvements in functional efficiency and the elimination of defects over the whole life of the building. It also emphasises the benefits to repeat clients of long-term, partnering relationships with all key suppliers.

Modernising Construction

ISBN 0-10-276901-X
A report by the Comptroller and Auditor General of the National Audit Office (NAO) and available from any Stationery Office bookshop or by contacting NAO. Tel: 020 7798 7400 Email: enquiries@nao.gsi.gov.uk Website: www.nao. gov.uk/publications

A comprehensive report which sets out in detail the many barriers to improving construction industry performance and describes the various industry initiatives since 1994. It concludes that better value means better whole-life performance and this can only come from total integration of the design and construction supply chain through a single point of contact. This ensures the involvement of the specialist suppliers in design from the outset, which is key to the elimination of inefficiency and waste, the achievement of optimum whole-life costs and the delivery of maximum functionality.

Local Government Task Force (LGTF) Rethinking Construction Toolkit

Available from the Customer Sales Department, Thomas Telford Ltd, Unit 1/K Paddock Wood Distribution Centre, Paddock Wood, Kent, TN12 6UU. Tel: 020 7665 2464 Fax: 020 7665 2245 Email: orders@thomastelford.com

This provides local authorities with a valuable support to the abandonment of outdated procurement practices that cause waste, in terms of the inefficient use of labour and materials, poor whole-life performance and poor functionality. It provides simple, practical 'How To' guidance that will enable local authority staff to introduce 'smart' procurement as recommended by the Egan Report.

For the key Canadian and Australian reports relating to best practice in construction procurement:

Achieving Excellence in Construction

Available from the Canadian Government website: www. innovationstrategy.gc.ca/cmb/innovation.nsf/SectorReports/ Construction

Produced by the Canadian Construction Research Board in response to the Federal Government's announcement on Canada's innovation strategy in June 2002. The section dealing with the challenges facing the construction sector is remarkably similar to that set out in reports in other developed countries such as the UK (Latham and Egan reports). The report then underlines this international commonality in the section on the international drivers for change. The report also differentiates between efficiency and productivity, and focuses on the industry's inefficiency as the main area for improvement.

Building and Construction Industries Supply Chain Project (Domestic)

Available from the Australian Department of Industry, Tourism and Resources website: www.industry.gov.au

Produced for the Department of Industry, Tourism and Resources in response to the Australian Government's Building and Construction Action Agenda in June 2001 because it was believed that supply chain management was the key to radical improvement of the Australian construction industry. The report sets out the failings of the Australian construction industry and explains how the importation of supply chain management from other business sectors could improve the performance of the construction industry. The report also looks at developments in other countries, but paints a fairly bleak picture of the rate of improvement in those other countries.

For detailed guidance on supply chain integration and management:
Building Down Barriers Handbook of Supply Chain Management – 'The Essentials'
ISBN 0-86017-546-4
Available from the Construction Industry Research and Information Association (CIRIA), 6 Storey's Gate, Westminster, London, SW1P 3AU.

An overview of the Building Down Barriers approach to supply chain integration and an introduction to the toolset as a whole. It describes the seven underlying principles of total supply chain integration and the lessons learned from their application on the two test-bed pilot projects. It also describes the benefits and the challenges of supply chain integration for the various sectors of the industry. The UK champion of the *Building Down Barriers Handbook of Supply Chain Management* is the Be organisation. The Be organisation also maintains the detailed and comprehensive toolset behind the handbook. Details of how to contact BE are provided later in this chapter.

For guidance on predicting and validating whole-life costs:
Whole Life Costing – A Client's Guide
Available from the Confederation of Construction Clients, 1ˢᵗ Floor, Maple House, 149 Tottenham Court Road, London, W1T 7NF. Tel: 020 7554 5340 Fax: 020 7554 5345 Email: cccreception@ccc-uk.co.uk Website: www.clientsuccess.org

A short, lucid guide (nine pages of text) that explains to clients the benefits, in business planning terms, of making construction investment decisions on the predicted cost of ownership. It explains the level of accuracy that can be expected at the various stages of design and construction and makes clear that optimum whole-life costs can only be achieved with the early involvement of specialist suppliers in design.

Technical Audit of Building and Component Methodology
Available from the Building Performance Group, Grosvenor House, 141–143 Drury Lane, London, WC2B 5TS. Tel: 020 7240 8070

Describes a technical audit process for assessing the whole-life performance of buildings and can be used as a first, second or third party audit system.

Housing Association Property Mutual (HAPM) Component Life Manual
Available from E and F N Spon, Cheriton House, North Way, Andover, Hampshire, SP1O 5BE. Tel: 01264 342933

The manual schedules over 500 components and gives the insured life, maintenance requirements and adjustment factors. The insured lives are cautious, were developed for housing, and are limited to 35 years, so should not be used without adjustment. The manual is updated twice a year, contains references to current British and European Standards and includes feedback from research and claims on HAPM latent defect insurance.

Building Services Component Life Manual – Building Lifeplans
Available from Blackwell Publishing Ltd, 9600 Garsington Rd, Oxford, OX4 2DQ. Tel: 01865 776868 Fax 01865 714591

This manual provides much needed guidance on the longevity and maintenance requirements of mechanical and electrical plant. It sets out typical lifespans of building service components – boilers, pipes, ventilating systems, hydraulic lifts, etc. These are ranked according to recognised benchmarks of specification, together with adjustment factors for differing environments, use patterns and operating regimes. Summaries of typical inspection and maintenance requirements are provided, along with specification guidance and references to further sources of information.

For detailed guidance on involving specialist suppliers (trades and specialist contractors) in design:
Unlocking Specialist Potential
ISBN 1-902266-00-5
Available from Reading Construction Forum, PO Box 219, Whiteknights, Reading, Berkshire, RG6 6AW.

A detailed guide that explains how the skill and experience of specialist suppliers can be harnessed in design development. It proposes strategies for better teamwork and collaboration, for a process-oriented approach to design and construction, and for a central focus on customer requirements. It makes clear that it is only by enabling the specialist suppliers to play a key role within the design process that real improvements in value can be achieved. The guide was used to develop the technology cluster concept in the Building Down Barriers toolset.

For detailed international evidence of labour inefficiency levels:
BSRIA Technical Note 14/97 Improving M & E Site Productivity
Available from the Building Services Research and Information Association, Old Bracknell Lane West, Bracknell, Berkshire, RG12 7AH. Tel: 01344 426511 Fax: 01344 487575 Email: bookshop@bsria.co.uk Website: www.bsria.co.uk

Comprehensive evidence from projects in UK, USA, Germany, France and Sweden on the true level of the efficient use of labour in mechanical and electrical services. It also describes the causes of the inefficiencies described in the report, including naming the sector of the industry that was responsible for causing the individual problem. It also gives advice on how to improve efficiency levels by better integration and co-ordination.

BSRIA Technical Note 13/2002 Site Productivity – 2002, A guide to the uptake of improvements
Available from the Building Services Research and Information Association, Old Bracknell Lane West, Bracknell, Berkshire, RG12 7AH. Tel: 01344 426511 Fax: 01344 487575 Email: bookshop@bsria.co.uk Website: www.bsria.co.uk

A comprehensive review of what has happened in the UK building services industry since the publication of TN 14/97, including detailed feedback from the small number of projects where the recommendations from TN 14/97 have been applied.

For an analysis of the key reports on the UK construction industry since 1944:
Construction Reports 1944–98
ISBN 0-632-05928-1
Available from Blackwell Publishing Ltd, 9600 Garsington Rd, Oxford, OX4 2DQ. Tel: 01865 776868 Fax: 01865 714591
A detailed analysis of the key reports on the UK construction industry since 1944, starting with the Simon Committee report in 1944 and ending with the Egan report in 1998. In its conclusion it picks up the recurring themes that run through the reports and thus demonstrates how little the industry's structure, culture and performance has improved since 1944.

For measurement of effective labour utilisation:
CALIBRE The Productivity Toolkit
For further information contact the Centre for Performance Improvements in Construction (CPIC), BRE, Garston, Watford, Hertfordshire, WD2 7JR.
CALIBRE provides a consistent and reliable way of identifying how much time is being spent on activities that directly add value to the construction and how much time is being spent on non-added value activities.

For training and coaching in construction and procurement best practice:
ICOM/CITB Diploma in Construction Process Management
For further information contact ICOM, Long Grove House, Seer Green, Buckinghamshire, HP9 2UL. Tel: 01494 675921 Fax: 01494 675126 Email: Kinder.ICOM@btinternet.com
ICOM is linked with the Construction Industry Training Board (CITB) and the University of Cambridge Local Examinations Syn-

dicate and uses the Construction Best Practice Programme booklet *A Guide to Best Practice in Construction Procurement* to
define best practice in its training. ICOM is also working with the
CITB to offer awareness workshops and coaching for clients on
best practice procurement. ICOM is also working with CITB to
develop coaching and training in best practice construction for
small and medium-sized construction industry firms (design consultants, construction contractors, specialist/trades contractors
and manufacturers).

For best practice techniques in managing the supply chain in the UK aerospace industry:

Sigma Management Development Ltd SCRIA Handbook
'Working Together'
Available from Sigma Management Development Ltd, 51
Peach Street, Wokingham, Berkshire, RG40 1XP. Tel: 0118
977 1855 Fax: 0118 977 4995
Email: sigma.mdl@btinternet.com

   Sigma Management Development Ltd, working in close collaboration with the UK Department of Trade and Industry and the
UK aerospace industry, developed the SCRIA Handbook in the
mid-1990s. The handbook forms the basis of SCRIA (Supply
Chain Relationships in Aerospace) training and leads to a
SCRIA qualification that is recognised across the UK aerospace
industry. The objectives of the SCRIA training are to help
members of the UK aerospace industry to achieve a positive
step change in performance through best practice in supply
chain management and to learn how to establish long-term, strategic supply-side partnerships within the supply chain for mutual
profit and successful growth through the elimination of unnecessary costs.

For advice on finding and working with integrated design and construction teams, and for collaborative forms of contract:

Be (previously the Design Build Foundation and Reading
Construction Forum)
PO Box 2874, London Road, Reading, RG1 5UQ. Tel: 0870
922 0034 Fax: 0118 975 0404 Website: www.beonline.co.uk

The Be organisation uniquely brings together representatives from the whole construction industry to champion the total integration of design and construction in order to deliver customer satisfaction through a single source of responsibility. It was formed from the merging of the Design Build Foundation and Reading Construction Forum in October 2002 and is a self-funded, multi-disciplinary organisation comprising leading construction industry clients, designers, consultants, construction contractors, specialist contractors, trades contractors, manufacturers and advisors. Be is the UK's largest independent organisation for companies across the whole design and construction supply chain.

Index